T0329522

Disturbance Observer for Advanced Motion Control with
MATLAB/Simulink

Disturbance Observer for Advanced Motion Control with MATLAB/Simulink

Akira Shimada
Shibaura Institute of Technology, Tokyo, Japan

IEEE Press Series on Control Systems Theory and Applications
Maria Domenica Di Benedetto, Series Editor

IEEE PRESS

WILEY

Published by John Wiley & Sons, Inc., Hoboken, New Jersey.
Published simultaneously in Canada.

For general information on our other products and services or for technical support, please contact our Customer Care Department within the United States at (800) 762-2974, outside the United States at (317) 572-3993 or fax (317) 572-4002.

Wiley also publishes its books in a variety of electronic formats. Some content that appears in print may not be available in electronic formats. For more information about Wiley products, visit our web site at www.wiley.com.

Library of Congress Cataloging-in-Publication Data Applied for:

Hardback ISBN: 9781394178100

Cover Design: Wiley
Cover Image: © Teera Konakan/Getty Images; © Pobytov/Getty Images

Set in 9.5/12.5pt STIXTwoText by Straive, Chennai, India

Author's Note

It is our great pleasure to have a chance to publish the book "Disturbance Observer for Advanced Motion Control with MATLAB/Simulink" and introduce it to readers worldwide.

Disturbance observer (DOB) is an algorithm or a function for estimating disturbances well-known to control engineers. Still, no book has been published that systematically and comprehensively explains its contents. For example, we can represent DOB with a transfer function or in a state-space representation.

Moreover, we can also design it as a digital system or use it in vibration or communication delay systems. Furthermore, we should consider the effects of noise and modeling errors when designing the system. This book includes all such problems and explains how to understand and handle these issues in an easy-to-understand manner with many examples using MATLAB/Simulink.

We initially published this book in Japanese in Autumn 2021 since we wanted to post it for Japanese control engineers and students. However, after the publishing, many friends, professors, and engineers recommended that I should publish it in English for engineers and students worldwide. This publication is a response to their strong encouragement.

The contents of this English version are the same as those of the Japanese version, but I revised all programs and the Simulink model using the MATLAB/Simulink R2022a version to make them clear for readers. The readers can download all sample programs from Wiley's home page.

Moreover, I initially selected many references written in Japanese. However, they are not convenient for readers worldwide. Therefore I reselected books and articles written in English.

We sincerely hope this book becomes helpful for you.

Books in the IEEE Press Series on Control Systems Theory and Applications

Series Editor: Maria Domenica Di Benedetto, University of l'Aquila, Italy

The series publishes monographs, edited volumes, and textbooks which are geared for control scientists and engineers, as well as those working in various areas of applied mathematics such as optimization, game theory, and operations

1. *Autonomous Road Vehicle Path Planning and Tracking Control*
   Levent Güvenç, Bilin Aksun-Güvenç, Sheng Zhu, Sükrü Yaren Gelbal

2. *Embedded Control for Mobile Robotic Applications*
   Leena Vachhani, Pranjal Vyas, and Arunkumar G. K.

3. *Merging Optimization and Control in Power Systems: Physical and Cyber Restrictions in Distributed Frequency Control and Beyond*
   Feng Liu, Zhaojian Wang, Changhong Zhao, and Peng Yang

4. *Model-Based Reinforcement Learning: From Data to Continuous Actions with a Python-based Toolbox*
   Milad Farsi and Jun Liu.

5. *Disturbance Observer for Advanced Motion Control with MATLAB/Simulink*
   Akira Shimada

Contents

About the Author

Dr. Akira Shimada was born in Chiba, Japan, in 1958. He received B.S. degree in electronics engineering from the University of Electro-Communications, Japan, in 1983 and received PhD in engineering from Keio University, Japan, in 1996. After graduation, he worked as a robotics engineer at Seiko Instruments from 1983 to 2001. He developed some industrial robot controllers. His main contribution was to create digital servo control systems, including disturbance observers, and design DC and AC servo motor drivers. Concurrently, he was a guest professor at Chiba University. He was an associate professor at the Polytechnic University, Japan, from 2001 to 2009. He has been a full professor at Shibaura Institute of Technology, Japan, since 2009. His current interests include motion control, robotics, control engineering, and free climbing. The present study themes are motion control and path planning and collision avoidance for humanoid climbing robots, wheeled mobile robots, inverted pendulum robots, autonomous drones, etc. His philosophy for the study is to have actual and practical experience. When he studies cooking robots, he cooks a variety of foods. To develop climbing robots, he climbs existing walls in the mountains or climbing gyms every week. He is a member of IEEE, SICE, and RSJ and a senior member of IEEJ.

Preface

There is an estimation method called the "disturbance observer." When some mechatronic systems move, friction, gravity, and external forces may disturb their motion. We refer to them collectively as disturbances. The disturbances are often unmeasurable, and the contents are unknown. If we can estimate the total value of these disturbances, we can improve the stability and tracking performance of the control system or use them in information processing. Generally, the term "observer" refers to a conference observer, but it is translated as "state observer" in control engineering and means a function for estimating state variables. The observer was proposed in 1964 by D. G. Luenberger, said to be a doctoral student at Stanford University then [1, 2]. The disturbance observer is an observer that estimates disturbances. Since the publication of the papers by Kiyoshi Ohishi, Kouhei Onishi, et al. [3] and Kouhei Ohnishi and Toshiyuki Murakami [4], they have attracted widespread attention and have been studied and applied by researchers and engineers worldwide [5–9]. Meditch and Hostetter [10] reported the design of a 0-observer for estimating unknown stationary inputs and a k-observer for estimating unknown inputs represented by k-degree polynomials. The extended system for observer design is defined using the unknown input $u(t)$ as $\bar{x} = [x^T(t), u^T(t)]^T$ in addition to the original state variable $x(t)$. Both $x(t)$ and $u(t)$ can be estimated according to the general observer theory if observability holds. This is not different from how we design disturbance observers in today's state space representation. Additionally, many papers have reported various unknown input estimation methods [11–16]. However, the name "disturbance observer" was invented by focusing intensely on the disturbance. Overall, "disturbance observer" includes almost complete disturbance suppression control by feeding back the disturbance estimate to cancel the disturbance, suppress parameter fluctuation, and control acceleration. It means that the "disturbance observer technology" is considered to have started with the papers [3, 4], and many others. This book aims to systematically describe the design process, application methods, and various properties of "disturbance observers" so that they can be helpful to many people who study control. In the design of disturbance observers for mechatronics system control, it is necessary to observe or calculate the position (or angle) or velocity (or angular velocity) information using sensors. It is essential to obtain highly accurate velocity information not affected by noise. We express our sincere thanks to Prof. Toshiaki Tsuji of Saitama University and Mr. Hiroyuki Nagatomi from Ohnishi Lab. They cooperated in writing the paper on velocity measurement and estimation techniques. The contents of this paper were tested in undergraduate and graduate classes, and then many suggestions were given by the members of the Motion Control Laboratory (Shimada Laboratory), especially Mr. Kenta Matsuo, Mr. Kazuki Tokushige, Mr. Katsumichi Takase, Mr. Ryoya Nakajima, and Ms. Yuka Kimura. In addition, Prof. Takashi Ohhira of Chuo University pointed out inadequacies in the descriptions and made many suggestions. Corona Publishing Co., Ltd. published this book as one of the new books solicited by the Society of Instrument and Control Engineers (SICE) in 2021. We express our sincere gratitude to Prof. Shiro Masuda of Tokyo Metropolitan University who was in charge of this book and the Publication Committee for their support in its completion. Finally, we thank Prof. Kouhei Ohnishi and many friends for their continual meetings and guidance.

References

1 David G. Luenberger: Observing the state of a linear system, IEEE Transactions on Military Electronica, Vol. 8, No. 2, 74–80, 1964.

2 George Ellis: Observers in Control Systems: A Practical Guide, Academic Press, 2014.

3 Kiyoshi Ohishi, Kouhei Ohnishi, Kunio Miyachi: Torque-Speed Regulation of Motor Based on Load Torque Estimation Method (IPEC-Tokyo '83), 1209–1218, 1983.

4 Kouhei Ohnish, Toshiyuki Murakami: Advanced motion control in robotics, 15th Annual Conference of IEEE Industrial Electronics Society (IECON '89), 356–359, 1989.

5 Asif Sabanovic, Kouhei Ohnishi: Motion Control Systems, Wiley-IEEE Press, 2011.

6 Emre Sariyildiz, Roberto Oboe, Kouhei Ohnishi: Disturbance observer-based robust control and its applications: 35th anniversary overvew, IEEE Transactions on Industrial Electronics, Vol. 67, No. 3, 2024–2053, 2020.

7 Shihua Li, Jun Yang, Wen-Hua Chen, Xisong Chen: Disturbance Observer-Based Control, CRC-Press, 2014.

8 Akita Shimada, Kiyoshi Ohishi, Masaaki Shibata, Osamu Ichikawa: EE text motion control, IEEJ & Ohmsha, 2004, 118–119, 157–164, 192–202 (In Japanese).

9 Akira Shimada: Recent advances and outlook in industrial instrumentation and MECHATRONICS control, IEEJ Transactions of Electrical and Electronic Engineering, Vol. 11, No. 52, S100–S107, 2016.

10 J.S. Meditch, G.H. Hostetter: Observers for systems with unknown and inaccessable inputs, International Journal of Control, Vol. 19, No. 3, 473–480, 1974.

11 Shih-Ho Wang, E. Journal Davison, Peter Dorato: Observing the states of systems with unmeasurable disturbances, IEEE Transactions on Automatic Control, Vol. 20, No. 5, 716–717, 1975.

12 C.D. Johnson: Optimal control of the linear regulator with constant disturbances, IEEE Transactions on Automatic Control, Vol. 13, No. 4, 416–421, 1968.

13 John O'Reilly: Minimal-order observers for linear multivariable systems with measurable disturbances, International Journal of Control, Vol. 28, No. 5, 743–751, 1978.

14 Tsutomu Mita: On the synthesis of an unknown input observer for a class of multi-input/output systems, International Journal of Control, Vol. 26, No. 6, 841–851, 1977.

15 Nobuaki Kobayashi, Takayoshi Nakamizo: An observer design for linear systems with unknown inputs, International Journal of Control, Vol. 35, No. 4, 605–619, 1982.

16 S.P. Bhattacharyya: Observer design for linear systems with unknown inputs, IEEE Transactions on Automatic Control, Vol. 23, No. 3, 483–484, 1978.

About the Companion Website

This book is accompanied by a companion website:

www.wiley.com/go/disturbanceobserver

The website includes sample programs with MATLAB/Simulink.

1

Introduction of Disturbance Observer

The **disturbance observer** is called as the pronoun of **motion control** and has been highly evaluated worldwide [1, 2]. Disturbances can be added to the input of the control plant, the output, or even any part of the **internal state**. There are two types of disturbance observers (DOBs): those that only estimate disturbances and those that also estimate state variables, such as position and velocity.[1] They are collectively called DOBs unless one wants to emphasize something in particular.

No matter how good the control or estimation method is, it is not a panacea. When a new control method is proposed, it is overhyped, leading to a boom. Examples include H_∞ control, sliding mode control, and model predictive control. However, as the boom continues for a while, their disadvantages become apparent, such as the need for the skill and technique in using them and their compatibility with the control plant. DOBs are no exception. It is essential to understand their pros and cons to use them well.

The intended readers are students and professionals learning the theories and techniques related to control engineering and motion control. They are assumed to have some knowledge of **classical control theory** and **modern control theory**. A minimal explanation is provided in the appendix for readers without sufficient knowledge.

1.1 Types of Disturbance Observers

1.1.1 Introduction

This book introduces eight types of DOBs for designing a DOB, as shown in Table 1.1. **Kalman filter** in the eighth line is not an observer but is included in

---

1 Examples have been proposed such as referring to it as disturbance and velocity estimation observer [3, 4] because it estimates velocity as well or reaction force estimation observer [5] because it estimates reaction force specifically.

Disturbance Observer for Advanced Motion Control with MATLAB/Simulink, First Edition. Akira Shimada.
© 2023 The Institute of Electrical and Electronics Engineers, Inc. Published 2023 by John Wiley & Sons, Inc.
Companion website: www.wiley.com/go/disturbanceobserver

Table 1.1 Types of disturbance observer design.

Systems	DOB design forms	Object to estimate
Continuous system	(1) Transfer function	Only disturbances
	(2) Identity observer	All state variables and disturbances
	(3) Minimal order observer	All state variables except outputs
	(4) Adaptive observer	Parameters, state variables and disturbances
Digital system	(5) Transfer function	Only disturbances
	(6) Identity observer	All state variables and disturbances
	(7) Minimal order observer	All state variables except outputs
	(8) Kalman filter	All state variables and disturbances

this table because it is designed for estimating disturbances. The Kalman filter and the adaptive observer can be designed as both continuous and digital systems, but they are limited to the above.

Many references use the **transfer function** of the continuous system to represent DOBs, specializing only in its estimation, which corresponds to the first line of Table 1.1. This may be because it is simple and easy for us to understand and implement. However, we can simultaneously estimate the control plant's original state variables, such as velocity and current other than disturbances, using the general observer theory. These correspond to lines 2–4 and 6–8 of the table.

Suppose that readers want to use various physical estimates effectively in designing the control system. In that case, they can design the DOB using the **identity observer** designing method. However, if they emphasize that they do not need to estimate the observed output, they can use the **minimal order observer** design method as an excellent choice.

Theoretically, DOBs expressed in transfer functions are equivalent to those derived from the design process for minimal order observers, as shown in rows 3 and 7. Many studies use the form of the transfer function from the beginning because only the disturbance estimation function is extracted and reexpressed in the form of a transfer function in the design process.

The difference between a continuous and a digital system is whether the design is based on continuous control theory or digital control theory.[2] If there is no need to consider the effect of the length or shortness of the control period, it is better to design a continuous system where the physical meaning is easy to grasp. However, it is more desirable to design it as a digital system if the control cycle and the

---

2 Note that this book does not distinguish between the words discrete systems and digital systems.

program to be implemented on a digital computer are considered. The best design method cannot be generally determined and is left to the designer's discretion.

1.1.2 Observer and Control System Design Concepts

The disturbance estimate $\hat{d}(t)$ in the DOB is often fed back with a sign for canceling disturbance.[3]

Figure 1.1 represents the structure of a basic control system that uses a DOB. In the figure, the "disturbance observer" outputs the disturbance estimate $\hat{d}$, and **positive feedback** is performed to cancel the disturbance d. Consequently, the control plant appears free of disturbances when viewed from outside the dotted section, and the apparent control input is $\bar{u}$. The control input is u, and $\bar{u}$ stands for the force f [N] for a linear mechatronics system, the torque τ [N m] for a rotating mechatronics system, and V [V] for the voltage input of an electric circuit. The output y is also chosen as the velocity v [m/s], ω [rad/s], position x [m], θ [rad], etc. In this book, we mainly use linear motion mechatronics systems as examples; thus, the explanation will be based on that assumption. For example, the equation of motion of a quality point in a vacuum is $f = ma$.

DOBs are often used to realize a control method called "**acceleration control.**" The nominal value of the mass m_0 is connected in a series in front of the dotted line, and the apparent control input is the acceleration reference value a_{ref}.[4]

Suppose the actual acceleration can be made to match the acceleration reference value. In that case, the control plant can be regarded as a single-dotted line, and the position and velocity controllers or the force control controller can be

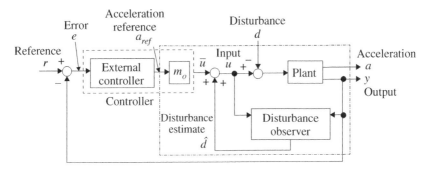

Figure 1.1 Basic structure of the control system based on acceleration control.

---

3 If the disturbance is input as a negative component, positive feedback is used to cancel it out (positive feedback), and negative feedback is used to define it with a positive sign (negative feedback). However, feedback is not always a must.
4 In this book, m denotes the mass of the control plant generally. However, only when we want to emphasize that it is a nominal value this book uses m_0.

implemented outside. This means that this could enable the design of a simple control system. The acceleration controller is configured inside the control system based on the concept of "moving a mechatronics system by acceleration," and the controller for each application is configured on the outside.[5]

Suppose a control system is to be designed based on the basic principle of dynamics that "a mechatronics system is driven by force (or a rotating system is driven by torque)." It is natural to use u or $\bar{u}$ (=force f for a linear motion system or torque τ for a rotating system) as the control input instead of a_{ref} and design the control system according to the general dynamics model. Here, m_0 is unnecessary because the part in the broken line represents the outer controller.

An alternative is to work without the feedback from the disturbance estimate $\hat{d}$. For example, when designing the identity DOB based on the state-space model, the purpose is not to estimate the disturbance but to estimate the state variable $x(t)$ with high accuracy considering the effect of the disturbance.

This book introduces the disturbance estimation method and various control methods using various concepts, including the case when the control plant is a **multiple input, multiple output (MIMO)** system.

We consider various conditions and environments such as nonlinearity, instability, and the effect of noise. Flexible use of classical control, modern control, and other robust control theories can further extend the usefulness of DOBs.

1.2 Format of Example and Use of MATLAB

1.2.1 Format of the Example Problem

Most studies dealing with DOBs are examples of rotating systems based on motor control. However, to understand the essential theory, this book introduces single-input single-output (SISO) system examples with a cart as an instance of a linear mechatronics system because we believe that it is simple and easy to understand.[6] The reader who needs to design a rotating system could convert the linear model to a rotating model.

Consider a simple spring-mass damper system. Using the position x, velocity v, force f, external force f_{ex}, **coefficient of viscous friction** c [N/(m/s)], and **coefficient of elasticity** k [N/m], the following equations can be obtained:

$$m\dot{v}(t) + cv(t) + kx(t) = f(t) - f_{ex}(t) \tag{1.1}$$

---

5 The concept of acceleration control is to contain the problems of the dynamic system inside. After that, it is to consider the kinematic model as the control plant. However, this does not mean that friction and the like have disappeared. Therefore, care must be taken not to overlook the essential aspects of mechatronics system structure and dynamics.
6 Some may argue that we are only dealing with low-dimensional objects, but we prioritize ease of understanding.

In contrast, assume that the rotation angle θ, rotation velocity ω, torque τ, disturbance torque τ_{ex}, moment of inertia of linear motion system J [kg m^2], coefficient of viscous friction c_θ [N m/(rad/s)], and the modulus of elasticity k_θ [N m/rad] satisfy:

$$J\dot{\omega}(t) + c_\theta\omega(t) + k_\theta\theta(t) = \tau(t) - \tau_{ex}(t) \tag{1.2}$$

Note that the disturbance estimation with the velocity information and that with the position information are often illustrated side by side.

1.2.2 Using MATLAB/Simulink

This book includes sample programs with "MathWorks" **MATLAB/Simulink®** to help the reader understand the specifics. Although the author is not an expert in MATLAB programming and does not have the latest knowledge, we have tried to make it accessible to a third party. However, the original m-file sample codes made by the author include long codes on how to draw figures as the simulation results, and the number of pages would be enormous. Hence, the sample m-files include up to the parts for calling and executing Simulink models but not the parts related to drawing figures. Readers would find the typical drawing program in the appendix helpful.

1.3 How This Book Is Organized

1.3.1 The Structure of This Document

The basics of DOBs are expressed in Chapter 2, mainly in the transfer function representation. Specifically, the concept of disturbances and the basic design methods are introduced, and disturbance rejection control and acceleration control methods are explained. An observer for estimating the reaction force is introduced, and then the **internal model principle** and **two-degrees-of-freedom** (2-DOF) **control system** are addressed. Finally, we describe the effect of modeling error.

In Chapter 3, we introduce the relationship between the stabilized control system using the coprime factorization and the control system with a DOB and show the design of the free parameter $Q(s)$ for the uncertainty of the control plant. This part is an encounter between the so-called "**robust control theory**" and the DOB.

Chapter 4 introduces DOBs and control system design methods for continuous systems based on modern control theory, especially **identity disturbance observer**, **minimal order disturbance observer**, **higher order disturbance observer**, and **periodic disturbance observer**. Additionally, observability, i.e. the possibility of observer design, is explained using a DC motor as a subject.

In Chapter 5, we present the design method for digital systems. As in Chapter 4, the same dimensions, minimum dimensions, and higher order disturbances are treated. We also explain the separation theorem, which allows us to design the control system's poles and the poles of the observer separately.

Chapter 6 deals with disturbance estimation for the **vibration system**. The disturbances in this chapter are neither input nor output disturbances but exactly disturbances to the internal state variables, i.e. **noninput/output disturbances**. However, they seem to be good examples of how they can be estimated if the observability is satisfied.

Chapter 7 introduces a technique to estimate the effect of **communication delays**, idle time disturbance, and stability maintenance.

Chapter 8 focuses on the **multirate control system**, which has recently attracted much attention and also introduces the impact of using a DOB in conjunction with it.

Chapter 9 introduces the **model predictive control** (MPC) combined with a DOB. Although the DOB in this chapter is not new and uses the observers in Chapter 5, we decided to introduce it because we believe the combination with MPC is meaningful.

In Chapter 10, we introduce a disturbance estimation method using the **Kalman filter**. The method is not a complete design method, and although the state variables can be estimated with high accuracy, the disturbance estimation is slow. To compensate for this shortcoming, the B and covariance matrices, subject to system noise, are set by trial and error. Other methods, such as assuming disturbances represented by higher order polynomials and modifying the error covariance matrix, are also reported, and future developments are expected.

In Chapter 11, we show that adaptive control theory can be used to design DOBs, such as the **adaptive disturbance observer**. An adaptive DOB is an observer that simultaneously estimates the parameters representing the control plant, the state variables, and the disturbances. This paper uses only classical design methods, and future developments are expected.

Chapter 12 is on velocity measurement and estimation, whose characteristics may feel different from all of the previous chapters. This chapter was added because there have been many reports on the use of velocity information in implementing DOBs. It is necessary to pay attention to the velocity measurement and estimation methods to ensure the accuracy of the estimation.

This book does not cover further developments in DOBs, such as **haptics** [6], or even applications to artificial intelligence. Nevertheless, we believe that this book will be found helpful and widely deployed.

1.3.2 How to Read This Book

While many technical books allow the reader to understand the book's core step by step as one reads from Chapter 1 onward, this book does not necessarily do so.

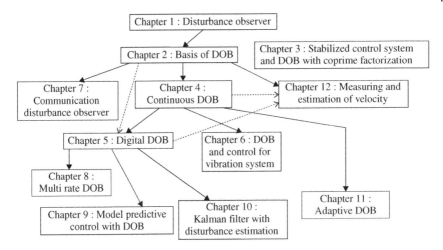

Figure 1.2 Structural diagram of this book.

Figure 1.2 shows the structure of the book. Here, the solid arrows indicate connection guides, and the dashed arrows are the candidates for connections. For instance, a reader who wants to learn about communication DOBs for systems with time delays can read Chapters 1, 2 and 7, skipping Chapters 3–6 even if the person does not know DOBs.

Chapters 8–10 are explained in digital systems and are recommended to be read after Chapter 5, but they are independent and do not need to be read in order. Furthermore, since many items are touched upon in Chapters 2, 4, and 5, some readers may not need to read through all of them. We hope that you will be able to "pick up" the items that interest you positively.

References

1 Kiyoshi Ohishi, Kouhei Ohnishi, Kunio Miyachi: Torque-Speed Regulation of Motor Based on Load Torque Estimation Method (IPEC-Tokyo '83), 1209–1218, 1983.

2 Kouhei Ohnish, Toshiyuki Murakami: Advanced Motion Control in Robotics, 15th Annual Conference of IEEE Industrial Electronics Society (IECON '89), 356–359, 1989.

3 Akira Shimada, Tsuyoshi Umeda, Norio Yokoshima, Naoki Kawawada, Hiroshi Watanabe, Toshimi Shioda, Masahide Nagai: Motion Control for Vertically Articulated Direct Drive Robot Manipulator, IEEE International Workshop on Advanced Motion Control (AMC'90), 132–137, 1990.

4 Yuhki Kosaka, Akira Shimada: Motion Control for Articulated Robots Based on Accurate Modeling, The 8th IEEE International Workshop on Advanced Motion Control (AMC'04), 535–540, 2004.

5 Toshiaki Murakami, Fangming Yu, Kouhei Ohnishi: Torque sensorless control in multidegree-of-freedom manipulator, IEEE Transactions Industrial Electronics, Vol. 40, No. 2, 259–265, 1993.

6 Kouhei Ohnishi, Seiichiro Katsura, Tomoyuki Shimono: Motion control for real-world haptics, IEEE Industrial Electronics Magagine, Vol. 4, No. 2, 16–19, 2010.

2

Basics of Disturbance Observer

What would be considered a disturbance? What mechanism is used to estimate it, and how are disturbance estimates handled? Classical control theory will be used to explain what happens in the presence of observation noise and modeling errors. In addition, the relationship between real systems equipped with DC motors and robust control theory will be explained.

2.1 What Is Disturbance

Let us define disturbance in this book.

> **Definition 2.1.1**   An input element that interferes with the control plant's motion except control input is called a **disturbance**. A disturbance that is applied to the input of the control plant is called an **input disturbance**; one that is applied to the control plant's output is called an **output disturbance**; and a disturbance that is applied to any other part of the control plant's interior is called a **noninput/output disturbance**.

Consider a mechatronics system; unexpected forces or torques may be generated inside the system due to estimation errors in physical parameters. In such cases, these forces and torques are interpreted as inputs that prevent movement from the outside and are collectively considered disturbances.[1]

Although the **noninput/output disturbance observer** estimates the disturbance inside the plant other than input and output, when we say disturbance observer generally, we almost mean the input disturbance observer. Thus, this manual mainly focuses on input disturbances.

1 We prefer to call external disturbances and internal disturbances caused by parameter errors, but in Japan they are collectively referred to as "external disturbances."

Disturbance Observer for Advanced Motion Control with MATLAB/Simulink, First Edition. Akira Shimada.
© 2023 The Institute of Electrical and Electronics Engineers, Inc. Published 2023 by John Wiley & Sons, Inc.
Companion website: www.wiley.com/go/disturbanceobserver

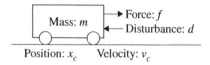

Position: x_c Velocity: v_c

Figure 2.1 Images of a linear motion mechatronics system (cart model).

However, in Section 4.3, an output disturbance observer is introduced, and an observer to estimate the output shaft disturbance of a two-inertia system is presented in Section 6 as an example of a noninput/output disturbance.

Generally, disturbances are often denoted by variables such as $d(t)$ and $d_{is}(t)$. The physical quantity of the input disturbance of a mechatronics system is the force or torque, or the input voltage or current to the drive circuit, and the output disturbance is the physical quantity of the position, velocity, force, temperature, etc. corresponding to the offset or output error of the sensor. As a simple example of an input disturbance, consider Figure 2.1. Given a mass m, velocity $v_c(t)$, force $f(t)$, and disturbance $d(t)$ that impedes the motion, the equation of motion is expressed by Equation (2.1):[2]

$$m\dot{v}_c(t) = f(t) - d(t) \tag{2.1}$$

If there is no frictional force, elastic force, or gravity, and only an external force f_{ex} acts as a disturbance to the cart's motion, then the external force f_{ex} is the disturbance; $d(t) = f_{ex}(t)$, and the equation of motion is given by Equation (2.2):

$$m\dot{v}_c(t) = f(t) - f_{ex}(t) \tag{2.2}$$

In Figure 2.2, we observe that on the cart we have **viscous friction** cv_c, elastic force kx_c, and **Coulomb friction** (kinetic friction) $f_c\mathrm{sign}(v_c)$. The equation of motion becomes Equation (2.3):

$$m\dot{v}_c(t) + cv_c(t) + f_c\,\mathrm{sign}(v_c(t)) + kx_c(t) = f(t) - f_{ex}(t) \tag{2.3}$$

If we consider only the external force f_{ex} as a disturbance, the disturbance remains $d(t) = f_{ex}(t)$, but we should also include the viscous frictional force cv_c, the elastic force kx_c, and the coulomb frictional force $f_c \cdot \mathrm{sign}(v_c)$ in the disturbance. If so, then the equation of motion is again expressed by Equation (2.3), and the disturbance

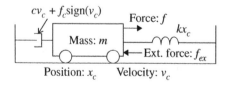

Position: x_c Velocity: v_c

Figure 2.2 Images of a linear motion mechatronics system (cart model) under reaction force.

2 $m\dot{v}_c(t) = f(t) + d(t)$, where d's value is negative if it impedes the motion. Note that the dot above the variable represents the time derivative d/dt, and the double dot represents d^2/dt^2.

by Equation (2.4). In the case of an acceleration control system, all elements other than the inertia term basically need to be defined as disturbances:[3]

$$d(t) = f_{ex}(t) + cv_c(t) + f_c \cdot \text{sign}(v_c(t)) + kx_c(t) \tag{2.4}$$

When an external force $f_{ex}(t) = 0$, and the cart is in contact with an elastic body and the reaction force $kx_c(t)$, the equation of motion and the disturbance are given by Equations (2.5) and (2.6):[4]

$$m\dot{v}_c(t) + cv_c(t) + f_c \, \text{sign}(v_c(t)) = f(t) - kx_c(t) \tag{2.5}$$

$$d(t) = kx_c(t) \tag{2.6}$$

What if the control target is a MIMO system? For example, consider a manipulator with multiple rotating joints. Let q be the joint angle vector, τ the driving torque, and the equations of motion be expressed as in Equation (2.7),

$$M(q)\ddot{q} + H(q, \dot{q}) + G(q) + \tau_{fric} = \tau - \tau_{ex} \tag{2.7}$$

where M is the inertia matrix, H is the nonlinear term, G is the gravity vector, τ_{fric} is the friction vector, and τ_{ex} is the external force vector. Let the disturbance vector d be $d = \tau_{ex}$. The decision to set d as $d = H(q, \dot{q}) + G(q) + \tau_{fric} + \tau_{ex}$ is the same as in the single-input single-output (SISO) system.

However, the MIMO system can be regarded as a collection of SISO systems, and an observer design can be performed. We can simplify the equations of motion using the nominal values of the moments of inertia at each joint and include the effects from different axes in the disturbance. For example, if we interpret τ_{ex} in Equation (1.2) to include the interference torque between the axes, we do not always need to design a MIMO system. Therefore, this book also focuses on the SISO system.

There is no rule regarding the control plant and what to consider as a disturbance. The designers should not be misled by the sound of the term "disturbance," but they should decide what physical quantity is to be estimated with considering the nature of the control plant.

2.2 How Disturbance Estimation Works

This section explains how the general and simple equation (2.1) and Figure 2.3 are used to estimate the disturbance of a cart. By applying the Laplace transform

---

3 $d(t) = cv_c(t) + f_c \cdot \text{sign}(v_c(t)) + kx_c(t)$; as mentioned above, it would be better to call it an "internal disturbance," but it is generally called an "Gairan (=disturbance)" without distinction between internal and external in Japan. In English, it is called internal disturbance.
4 The observer that estimates only the reaction force is called the **reaction force observer**. Some interpretations include viscous friction cv_c in the reaction force, and the elastic force is not necessarily linear kx_c but is a simple example.

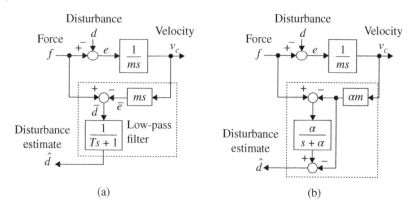

(a)   (b)

Figure 2.3 Basic principle and representation format of disturbance observer. (a) Basic block diagram and (b) format often used for implementation.

to the initial value of the cart's velocity $v_c(0)$, 0 m/s, we obtain $m \cdot (sv_c(s) - v_c(0)) = m \cdot sv_c(s) = f(s) - d(s)$, and the transfer function block diagram can be represented as shown in the upper part of Figure 2.3a. In this case, the deviation of the force e is set to $e = f - d$. The part in the dashed line in the figure represents the disturbance observer. The calculated value $\bar{e} = ms \cdot v_c = ms \cdot \frac{1}{ms}e = e$. Furthermore, $\bar{d} = f - \bar{e} = f - e = f - (f - d) = d$ holds. In other words, $\bar{d}$ is the calculated value of the disturbance. The estimated disturbance value $\hat{d}$ is the value obtained by passing the calculated disturbance value $\bar{d}$ through a low-pass filter with a time constant T to remove the high-frequency component on the force f, the observation noise of the velocity v_c, and the noise component generated by the differential operator in the ms section. However, rather than implementing the disturbance observer in the form shown in Figure 2.3a, it is often implemented in the form shown in Figure 2.3b, which excludes the derivative operation.

The process of transforming Figure 2.3a into b is described below.

1. $\frac{1}{Ts+1} = \frac{1/T}{s+1/T} = \frac{\alpha}{s+\alpha}$. That is, $\alpha = 1/T$.
2. The transfer function from v_c to the output of $\frac{1}{Ts+1}$ is $-\frac{ms}{Ts+1} = \frac{-\alpha ms}{s+\alpha}$, which contains the derivative s in the numerator. So, we expand it as follows

$$\frac{-\alpha ms}{s+\alpha} = \frac{x_1}{s+\alpha} + x_2 = \frac{x_2 s + x_1 + x_2 \alpha}{s+\alpha} \tag{2.8}$$

3. Since $x_2 = -\alpha m$ and $x_1 = -x_2\alpha = \alpha^2 m$ must be true, we get $\frac{-\alpha ms}{s+\alpha} = \frac{\alpha^2 m}{s+\alpha} - \alpha m = \alpha m(\frac{\alpha}{s+\alpha} - 1)$.

Applying this result to the basic block and separating out the αm, we get the implementation format. Despite being equivalent. It has the advantage of not requiring a derivative calculation.

Observers are generally designed in a state space representation, so readers who have learned the basics of modern control theory may feel uncomfortable calling a function that performs disturbance estimation in the form of a transfer function an observer. However, it is no problem because Equation (2.3) is equivalent to extracting the disturbance estimation part of the minimal order disturbance observer described below and reexpressing it in transfer function form.

2.3 Disturbance Rejection and Acceleration Control System

2.3.1 Concept of Disturbance Rejection and Acceleration

As a typical usage of the disturbance observer, the control input $f = \bar{f} + \hat{d}$ is often used, as shown in Figure 2.4a. In other words, $\hat{d}$ is used to cancel out the disturbance d. The $\bar{f}$ is interpreted as the apparent control input after canceling out the disturbance. If $\hat{d}$ converges to d exactly, then equivalently we get (b), and the part in the single-dotted line is organized as (c). The transfer function $s/(s + 1/T)$ has a high-pass filter property, and if we draw a bode diagram as shown

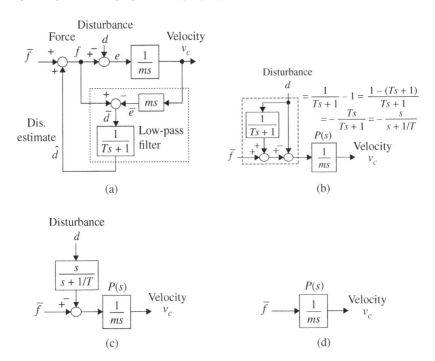

(a)

(b)

(c)

(d)

Figure 2.4 Principle of disturbance rejection using disturbance estimates. (a) Principle of disturbance rejection, (b) equivalent disturbance rejection (1), (c) equivalent disturbance rejection (2), and (d) apparent control plant.

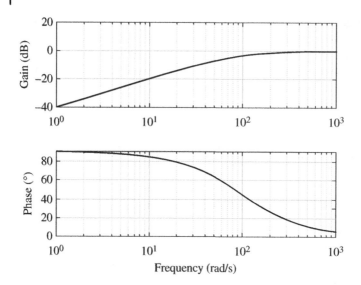

Figure 2.5 Board diagram for $G(s) = s/(s + 1/T)$ (example with $T = 0.01$ s).

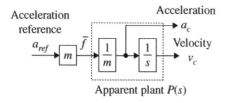

Apparent plant $P(s)$

Figure 2.6 Apparent acceleration control system.

in Figure 2.5, the frequency domain $\omega = 1/T\ (= \alpha)$. In this example, the disturbance equivalence factor is less than 0 dB in the region below 1/T(= 100 rad/s), which means that the disturbance is greatly attenuated in the low-frequency band, i.e. the disturbance valid can be removed. The disturbance at 1 rad/s becomes 1/100 (=−40 dB), and the disturbance at 10 rad/s will be 1/10 (=−20 dB). In other words, if we can design for a small enough T value, the control plant can be treated as an apparently disturbance-free control plant, as shown in Figure 2.4d.

Furthermore, as in Figure 2.6, if a gain corresponding to the mass m is inserted in the first stage of the apparent force $\bar{f}$ and the input is the acceleration reference a_{ref}, the output is the acceleration a_c, $a_c = a_{ref}$, and the ideal acceleration control system can be realized.[5]

_____

5 When considering the error between the mass m and the nominal value m_0, set the input block's gain to m_0 instead of m.

2.3.2   Different Disturbance Observers Depending on How the Disturbance Is Captured

In Sections 2.1 and 2.2, we introduced various cases where the object of disturbance is not only the external force f_{ex} but also the viscous friction, Coulomb friction, gravity, reaction force, etc. Using the example of external force $f_{ex}(t)$ and viscous friction $cv_c(t)$, we learn how to handle disturbances. Specifically, we compare the case where viscous friction is included in the model to be controlled and the case where it is included in the disturbance.

For Figure 2.7a, we set $P^{-1}(s) = ms + c$, found $e = f - d = \bar{e}$, and the disturbance estimate $\hat{d}(t)$ evaluated the external force $f_{ex}(t)$ as the disturbance $d(t)$. In Figure 2.7b, $\bar{e} = e$ holds, and $\hat{d}(t) = d(t) = f_{ex} + cv_c(t)$ is estimated. Both are disturbance observers; the difference arises only after returning the disturbance estimate $\hat{d}(t)$.

Let us examine the case of positive feedback on the disturbance estimate for Figure 2.7a with viscosity considered. The block diagram becomes Figure 2.8a, and by performing the same equivalent transformation as in Figure 2.4, the apparent control plant becomes Figure 2.8b. As in Figure 2.6, even if the mass m gain is placed before the input f and $f = ma_{ref}$, the a_{ref} does not become the acceleration reference value, which does not fit the idea of controlling the mechatronics system by **acceleration control**.

Figure 2.7b can be used to configure an acceleration control system. Note that if we adopt the method shown in Figure 2.7a and leave the viscous friction cv_c, the poles of the equivalent control plant become $(-c/m)$, which is stable. However, if we cancel the viscous friction as shown in Figure 2.7b, we have to destabilize the control system and stabilize it with the outer controller.

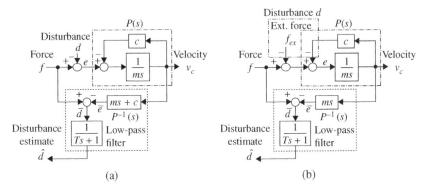

(a)   (b)

Figure 2.7   How to handle disturbances. (a) Viscous friction is included in plant and (b) viscous friction is included in disturbance.

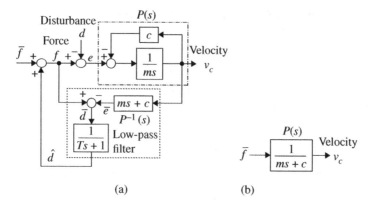

(a)   (b)

Figure 2.8 Disturbance cancel using disturbance estimates for a system with viscosity. (a) Disturbance rejection control and (b) equivalent system.

2.3.3 Basic Control System Design

Design the velocity control system after configuring the acceleration control system in Section 2.3.1.

Figure 2.9a is an example of a block diagram of a PI velocity control system that includes an acceleration control system. If the ideal acceleration control holds, we can regard $a = a_{ref}$ and interpret the control system in Figure 2.9a as the simple control system in Figure 2.9b. This transfer function can be summarized as $G_c(s)$ in Equation (2.9):

$$G_c(s) = \frac{\frac{K_p s + K_i}{s^2}}{1 + \frac{K_p s + K_i}{s^2}} = \frac{K_p s + K_i}{s^2 + K_p s + K_i} = \frac{2\zeta\omega_n s + \omega_n^2}{s^2 + 2\zeta\omega_n s + \omega_n^2} \tag{2.9}$$

This can be regarded as a second-order delay system and defined as $K_p = 2\zeta\omega_n$ and $K_i = \omega_n^2$. For example, the parameter values are determined as $\zeta = 1$, $\omega_n = 10 (\ll 1/T)$.[6] Alternatively, set the poles of the control system to p_1 and p_2, so that the real part is negative, and set the denominator polynomial to $s^2 + K_p s + K_i = (s - p_1)(s - p_2) = s^2 - (p_1 + p_2)s + p_1 p_2$. You may also use $K_p = -(p_1 + p_2)$ and $K_i = p_1 p_2$.

Examples of programs and simulations are shown in List 2.1 and Figure 2.10.

---

6 Unlike the basic second-order delay system $\frac{\omega_n^2}{s^2 + 2\zeta\omega_n s + \omega_n^2}$, the first-order term, s, in the numerator tends to cause overshoot, so we do not stick to the general critical braking $\zeta = 1$. We chose a value with a good response waveform in the following example.

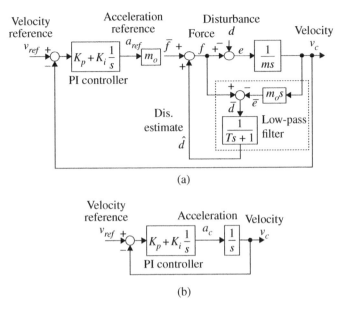

(a)

(b)

Figure 2.9 Block diagram of PI velocity control system with DOB. (a) Speed control system and (b) equivalent speed control system.

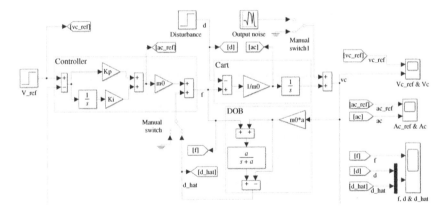

Figure 2.10 Simulink model of basic PI velocity control system with DOB.

List 2.1: PI velocity control system with DOB.

1 %% Physical parameters
2 m0=1; % Nominal value of mass [kg]
3 d=10; % Magnitude of disturbance [N]
4 T=0.01;a=1/T; % Time constant of DOB
5 zeta=7.5;wn=5; % Damping constant,Resonance frequency

```
6   Kp=2*zeta*wn;Ki=0; % wn^2, PI control gains
7   dt=0.005;d_out=0.001; % Period and magnitude of random noise
8
9   %% Simulation
10  tend=5; % Simulation time
11  open_system('sim_Figure_2_10_and_2_11_basic_vel_con_PI_DOB');
12  y=sim('sim_Figure_2_10_and_2_11_basic_vel_con_PI_DOB');
```

Assuming a nominal value of mass $m_0 = 1$ kg and a disturbance $d = 10$ N, $T = 0.01$ s, $\zeta = 7.5$, and $\omega_n = 5$ rad/s. The disturbance observer for Figure 2.10 is the type of Figure 2.3b.

Figure 2.11 shows the simulation results.

The figures of Columns (a) and (b) of Figure 2.11 shows the simulation results corresponding to the disturbance estimators' positive feedback ON and OFF, respectively.

Although the disturbance estimation functions are equal, the acceleration waveform in the middle row is almost identical to the reference value in Figure 2.11 (a-2) but deviates from it due to the influence of the disturbance in (b-2). The disturbance in (a-3) also suppresses the velocity waveform in the lower row, but the deviation is generated in (b-3).

2.4 Reaction Force Observer (RFOB)

2.4.1 Reaction Force Observer Design

When the control plant and output disturbance are defined in Equations (2.5) and (2.6), how can we estimate $f_{reac}(t) = kx_c(t)$ as the reaction force from the elastic wall. What should we do? If we consider the estimation principle in [1], we can draw Figure 2.12a. The f_c sign(v_c) and $cv_c + f_c$ sign(v_c) blocks should output the result of the calculation in the block, not the multiplication gain, omitted. However, if it cannot be ignored, the static friction model f_{static} must be derived and added [2, 3]. The dashed line is called **reaction force observer** (RFOB).

The estimation principle is explained below. The velocity v_c is time differentiated and multiplied by m to obtain the calculated value of the signal $h, \bar{h}$. From $h = e - \{cv_c + f_c$ sign(v_c)\}, we obtain the calculated value of $e, \bar{e} = \bar{h} + cv_c + f_c$ sign(v_c). The calculated reaction force $f_{reac} = f - e$ is obtained using the relation $f_{reac} = f - e$, and the reaction force estimate $\hat{f}_{reac}$ is obtained through the low-pass filter.

Alternatively, the configuration in (b), which is in line with Figure 2.3, can be considered, where $\tilde{f}_{reac}$, the reaction force estimate, can be used for force control instead of the force sensor [1, 4]. In other words, **force sensorless force control** is possible. Moreover, the reaction force need not be linear as kx_c in this example. Since it is very convenient, it has been used for flexible arm control, bilateral control, etc. [5, 6].

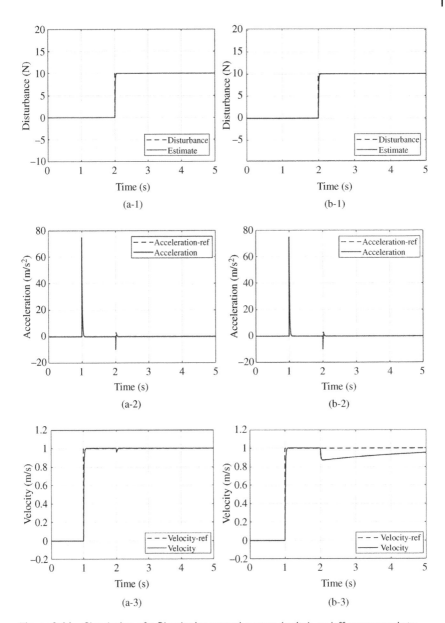

Figure 2.11 Simulation of a PI velocity control system (switch on/off corresponds to whether the disturbance estimate has a positive feedback or not). (a-1) Disturbance waveform with SW on, (a-2) acceleration waveform with SW on, (a-3) velocity waveform with SW on, (b-1) disturbance waveform with SW off, (b-2) acceleration waveform with SW off, and (b-3) velocity waveform with SW off.

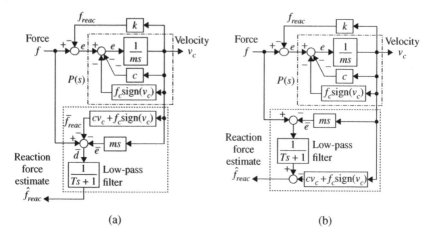

(a)   (b)

Figure 2.12 Basic principle of reaction force observer. (a) Basic reaction force estimation structure and (b) example with basic DOB.

2.4.2 Combined Use of DOB and RFOB

An example of how the RFOB realizes force sensorless control is shown in Figure 2.13. In the inner loop, the positive feedback of the $\hat{d}$ value is used to cancel the disturbance including the reaction force. Moreover, the outer loop is used to control the reaction force value again.

Following the idea of "moving a mechatronics system by acceleration" described earlier, let us design a reaction force control system with the single-dotted line as the apparent control plant and the apparent control input as a_{ref}.

Example 2.4.1 An example program for the cart model is shown in List 2.2; an example Simulink model is shown in Figure 2.14; and the simulation results are shown in Figure 2.15.

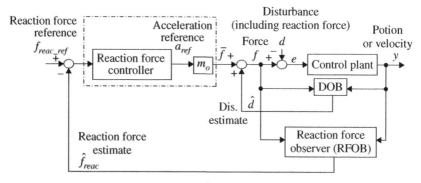

Figure 2.13 Block diagram of the combined DOB/RFOB reaction force control system.

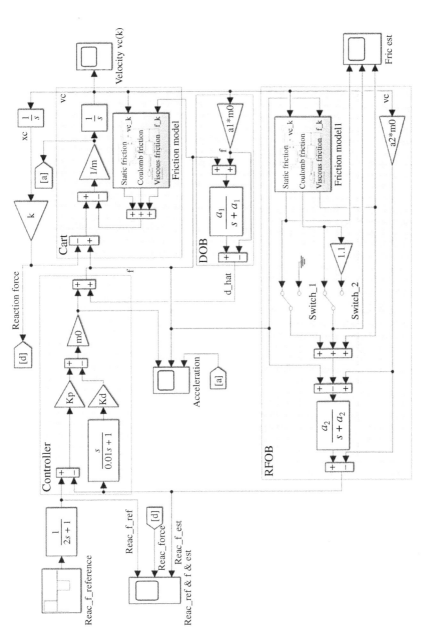

Figure 2.14 Simulink model example of Reaction force control with RFOB.

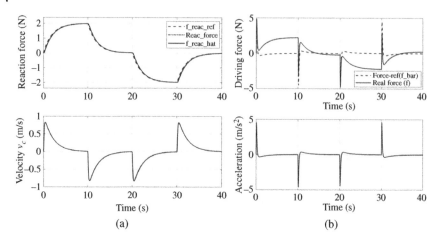

Figure 2.15 Simulation result of reaction force control with RFOB. (a) Reaction force and velocity and (b) force and acceleration.

List 2.2: Reaction force control with RFOB.

```
1   %% Physical parameters
2   m=1.0;m0=m;c=0.1;k=1; % Mass,viscosity,modulus of elasticity,
3   fric=0.25; % Magnitude of Coulomb friction
4   fs= 1; % Magnitude of maximum static friction
5   ref= 2; % Magnitude of reaction force reference
6   % The explanation for modeling static friction was omitted.
7   %
8   T=0.02; % Time constant of DOB
9   dt=0.001; % Maximum of time interval in Simulink
10  a1=1/T; a2=a1; % (-1)*pole of DOB, (-1)*pole of RFOB
11  delta=[1e-4,1e-3,1e-5,2e-3]; % Threshold parameters used
12  % in MATLAB function for calculation of friction.
13  % delta=[Velocity, acceleration,stop,Coulomb friction]
14  a3=12; % Damping parameter of Stribeck curve line
15  a4=1e-3; % Gain for Sigmoid function to express Coulomb friction
16  Kp=75;Kd=15; % Pd controller gains
17
18  %% Simulation
19  t_end=40; % Simulation time
20  open_system('sim_Figure_2_14_and_2_15_reaction_con_RFOB');
21  y=sim('sim_Figure_2_14_and_2_15_reaction_con_RFOB');
22
23  % Switch_1 is provided to switch between with and without static friction.
24  % Switch_2 is prepared to test the case where Coulomb friction is increased by 10%.
```

The friction model and the friction estimation model in Figure 2.14 are based on Iwasaki et al. The friction model and the friction estimation model in Figure 2.14 are MATLAB functions based on the mathematical expressions for static, Coulomb, and viscous frictions. The poles of both observers were set to $-a_1, -a_2 (= -1/T)$ and implemented in the form of Figure 2.3b. Additionally, the reaction force control controller was a PD controller. In Figure 2.15a, the reaction force (Reac force) and the estimated value (f_reac_hat) follow the reaction force reference value (f reac_ref).

The velocity waveform is shown in the figure below.[7]

In the figure above, the apparent driving force $\bar{f}$ is shown as a dashed line, the driving force (real force (=f)) is shown as a solid line, and the figure below shows the acceleration.

In the acceleration control, it is often expressed as "mapping force to acceleration," but care must be taken to determine which part of the force corresponds to the acceleration. In this example, as seen in Figure 2.15b, $\bar{f} = m_0 a_c$ is valid for the apparent force ($\bar{f}$), but $f = m_0 a_c$ is invalid for the actual force (real force in (b)). This is not the case.

As shown in Figure 2.14a, if we turn off Switch 1 in the figure and omit the static friction, the simulation waveform diverges, reducing the PD gain value. Alternatively, the reaction force control system became stable when the value of the maximum static friction (f_s value in List 2.2) was set to a small value. Next, when the Coulomb friction was estimated to be about 10% larger (gain 1.1), the reaction force waveform was generally normal, but oscillations occurred in the velocity waveform. Oscillations were also observed in the smaller case (gain of 0.9), but in both cases, the oscillations disappeared when the PD gain value was reduced. As shown above, even when the modeling error of friction is large, it may be possible to solve the problem by keeping the control bandwidth low. However, it is desirable to create a highly accurate friction model.

As described earlier, there is a difference in whether the reaction force is given as an external input or generated by the interaction with the environment. In the former case, the reaction force does not contribute to the stability, while in the latter case, the reaction force affects the stability of the system because the system is a combination of the mechatronics system and the environment.

The design of the environment model that generates the reaction force is important in the latter case. For example, if the environment is a rigid body, there are two possible methods: one is to estimate the modulus of elasticity as extremely large, and the other is to derive the equations of motion considering the

---

7 In force control experiments against a rigid wall, the robot often struggles with repeated contact and disengagement **hunting phenomenon**, but in this example, the robot is assumed to have a negative reaction force without disengagement.

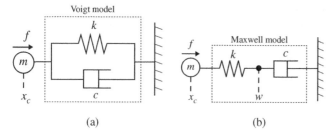

Figure 2.16 Two kinds of environment models. (a) Voigt model and (b) Maxwell model.

constraint conditions, including **Lagrange's unknown multiplier**. The former is an approximate model. In the latter method, a rigorous theoretical development is possible, but we omit it for the sake of space.

In this book, we have assumed that the environment that generates the reaction force is an elastic wall, but depending on the nature of the wall, it may not return to its original position even if the force is removed after pushing the wall. For example, it is the case that the environment is not the Voigt model in Figure 2.16a but the Maxwell model (b),[8] a generalized version of this model is called the **rheology model**, which is required for food grasp control and other applications [7].

2.5 Internal Model and Two-degrees-of-freedom Control

2.5.1 Internal Model Principle

Have you ever heard of the **internal model principle**? It is a very important principle in learning servo systems and is also essential in control systems with disturbance observers. However, many books do not clarify whether the principle is about the convergence value of the deviation $e(t)$ with respect to the reference value $r(t)$ or about the disturbance $d(t)$. In this book, it is defined as follows.

Definition 2.5.1 (Internal model principle.) Suppose that the Laplace transform of the reference $r(t)$ is expressed as $r(s) = g(s)r_0$. In this case, the behavior of the deviation $e(s) = r(s) - y(s)$ consists of a term related to the reference value $r(s)$ and a term related to the disturbance $d(s)$.

(continued)

---

8 The model in this book, with kx_c as the reaction force, belongs to (a).

(*Continued*)

1. The control plant must have the same poles as $g(s)$; otherwise, the controller must have the same poles as $g(s)$.
2. When the input end disturbance is expressed as $d(s) = h(s)d_0$, the controller $C(s)$ or the controller containing the disturbance observer feedback must also have the same poles as $h(s)$.
3. The controller or the controller containing the disturbance observer's feedback must also have the same poles as $h(s)$.

In a simple example, the transfer function of the step shape reference value is sometimes represented by $r(s) = 1/s \cdot r_0$ and the disturbance is represented by $d(s) = 1/s \cdot d_0$. Then an integrator $1/s$ should be included in the controller for the output to follow the reference value exactly.

If we have a disturbance observer with an internal model, we can find whether the deviation $e(t) = r(t) - y(t)$ can converge to 0 even for a reference value containing integrals of the same order.

Figure 2.17 is the basic structure of the control system with a disturbance observer.

Based on Figure 2.17, expressing the deviation $e(s)$ in terms of the reference value $r(s)$ and disturbance $d(s)$, we obtain Equation (2.10):

$$
\begin{aligned}
e(s) = & \frac{1 + H(PP_0^{-1} - 1)}{1 + P(s)C(s) + H(PP_0^{-1} - 1)} r(s) \\
& + \frac{P(s)(1 - H(s))}{1 + P(s)C(s) + H(PP_0^{-1} - 1)} d(s)
\end{aligned}
\tag{2.10}
$$

On the other hand, if there is no modeling error and $P(s) = P_0(s)$ holds, then $(PP_0^{-1} - 1)$ vanishes, and we can consider Equation (2.11). The case of $P_0(s) \neq P(s)$ and stability is treated in the following section.

$$
e(s) = \frac{1}{1 + P(s)C(s)} r(s) + \frac{P(s)(1 - H(s))}{1 + P(s)C(s)} d(s)
\tag{2.11}
$$

Figure 2.17 Basic structure of control system with DOB.

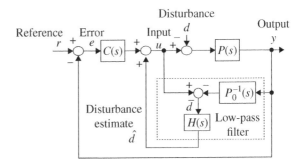

In Equation (2.11), the deviation $e(s)$ is divided into a term for the reference value $r(s)$ and a term for the disturbance $d(s)$. The first term on the right corresponds to the reference value response (also called the target value response), and the second term corresponds to the disturbance response. The disturbance response is determined by $H(s)$, which exists only in the second term, in addition to $P(s)$ and $C(s)$, so the properties of the reference response and the disturbance response can be designed separately. In the sense that there are two-degrees-of-freedom in the design, we say **two-degrees-of-freedom control** system.[9]

In other words, the disturbance observer is irrelevant to the response characteristics of the reference value, and whether or not the deviation can be set to a 0 value depends on the first term on the right-hand side. However, whether the deviation can be set to 0 for unknown disturbances depends on the second term on the right-hand side. The final value of the deviation can be proved using the **final value theorem**, and an example will be given later. By the way, Figure 2.17 can be transformed by splitting the disturbance observer part into two loops, Figure 2.18a,b. The structure of the $1/(1 - H(s))$ part in the figure confirms an essential property of the control system with a disturbance observer.

Example 2.5.1 The final value of a control system with first-order $H(s)$.
Let $P(s) = \frac{1}{ms}$ and $H(s) = \frac{1}{Ts+1}$. In addition, to make it easier to understand the internal model, consider a simple proportional control of $C(s) = K_p$, the controller of Figure 2.9. If $m_0 = m$, we get Equation (2.12):

$$e(s) = \frac{s}{s + K_p/m}r(s) + \frac{sT/m}{(Ts+1)(s+K_p/m)}d(s) \qquad (2.12)$$

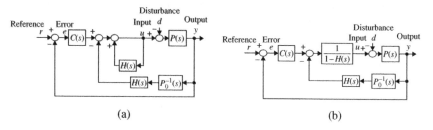

(a)   (b)

Figure 2.18 Equivalent transformation for control system with DOB. (a) Equivalent system 1 and (b) equivalent system 2.

---

9 The right-hand side of the equation $S(s) = \frac{e(s)}{r(s)} = \frac{1}{1+P(s)C(s)}$ is called the **sensitivity function**. Also, $T(s) = \frac{y(s)}{r(s)} = \frac{P(s)C(s)}{1+P(s)C(s)}$ is called the **complementary sensitivity function**, and $S(s) + T(s) = 1$ holds.

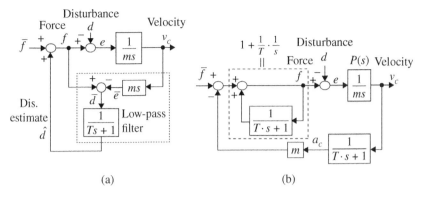

Figure 2.19 Equivalent transformation system with positive feedback of disturbance estimate. (a) Disturbance rejection control (again) and (b) equivalent transformations.

Assuming a step reference $r(s) = \frac{r_0}{s}$ and a step disturbance $d(s) = \frac{d_0}{s}$, and using the final value theorem, we obtain Equation (2.13):

$$e(\infty) = \lim_{s \to 0} se(s) = \lim_{s \to 0} \left\{ \frac{s}{s + K_p/m} \frac{r_0}{s} + \frac{sT/m}{(Ts + 1)(s + K_p/m)} \frac{d_0}{s} \right\}$$
$$= 0 \tag{2.13}$$

Meanwhile, if we use ramp inputs such as $r(s) = \frac{r_0}{s^2}$ and $d(s) = \frac{d_0}{s^2}$, then $e(\infty) \neq 0$. The calculation process is omitted.

Next, let us rewrite Figure 2.19a with positive feedback of the disturbance estimate as Figure 2.19b in two separate loops. The dashed part $1/(1 - H)$ is equivalent to the PI controller $1 + 1/(Ts)$; the internal model $1/s$ can be seen. In other words, the internal model is generated by the **positive feedback** of the disturbance estimate.

Example 2.5.2 The final value of the control system with second-order system $H(s)$.

If the control plant $P(s) = \frac{1}{ms}$ and $H(s) = \frac{g_1 s + g_2}{s^2 + g_1 s + g_2}$ ($g_1 > 0, g_2 > 0$).[10] Similarly, consider the proportional control of $C(s) = K_p$ by replacing PI with P for the controller of Figure 2.9a, and we obtain Equation (2.14):

$$e(s) = \frac{s}{s + K_p} r(s) + \frac{s^2}{m(s + K_p)(s^2 + g_1 s + g_2)} d(s) \tag{2.14}$$

---

10 $H(s) = \frac{g_2}{s^2 + g_1 s + g_2}$ is also considered. This is possible. In fact, we have adapted it to an example that is equivalent to the minimal order disturbance observer, which we will show later.

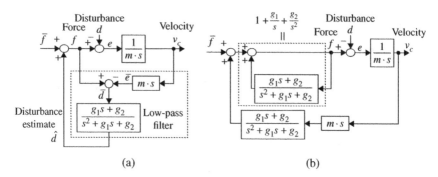

Figure 2.20 Positive feedback of first-order disturbance estimate. (a) Disturbance removal control system and (b) equivalent transformation.

Let us assume that we have a ramp reference $r(s) = \frac{r_0}{s^2}$ and **ramp disturbance** $d(s) = \frac{d_0}{s^2}$. The final value of the deviation $e(t)$ is

$$e(\infty) = \lim_{s \to 0} \left\{ \frac{s}{s + K_p/m} \frac{r_0}{s^2} + \frac{s^2}{m(s + K_p)(s^2 + g_1 s + g_2)} \frac{d_0}{s^2} \right\}$$

$$= \frac{r_0}{s + K_p/m} + 0 \neq 0 \tag{2.15}$$

In other words, positive feedback from the disturbance estimate $\hat{d}(t)$ eliminates the ramp disturbance but does not follow the ramp reference. In other words, it is not the so-called 2-type velocity servo system.

Next, let us look at the case of positive feedback of the first-order disturbance observer's disturbance estimate in the velocity observation. An equivalent transformation of Figure 2.20a diagram yields Figure 2.20b. It can then be observed that an internal model of type $1/s^2$ is constructed.[11]

2.5.2 Feedforward Control

Feedforward control uses inputs to produce a known behavior in advance, separate from the feedback control inputs. The feedback of the "disturbance observer" is sometimes described as "feedforward" because the idea is to predictably cancel out the disturbance separately from the main loop feedback.

The term "two-degrees-of-freedom control system" is used as a control system where the tracking characteristics and the design of the stability and

---

11 In the past, it has been questioned whether control using a disturbance observer is the same as mere integral control. This example shows that there are different properties.

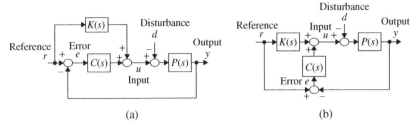

Figure 2.21 Structure of feed-forward control system. (a) Example of a feedforward control system1 and (b) example of a feedforward control system2.

disturbance rejection characteristics of the feedback control system are designed independently of the number of degrees of freedom of the mechatronics system's motion. Generally, a feedback control system without feedforward (i.e. one degree of freedom control system) cannot be designed independently of each other, but it can be designed independently by adding a feedforward control part $K(s)$, resulting in a two-degrees-of-freedom control system. Two examples of block diagrams for a control system including a typical feedforward function are shown in Figure 2.21. Note that both are entirely equivalent.

As in the previous section, we can find the deviation $e(s)$ with respect to the reference value $r(s)$ and the disturbance $d(s)$ to obtain Equation (2.16):

$$e(s) = \frac{1 - P(s)K(s)}{1 + P(s)C(s)}r(s) + \frac{P(s)}{1 + P(s)C(s)}d(s) \tag{2.16}$$

Looking at the first term on the right-hand side of Equation (2.16), we can see that if we set $K(s) = P^{-1}(s)$, we can set $e(s) = 0$ for $r(s)$, independent of $C(s)$. The second term on the right-hand side does not have $K(s)$ and $C(s)$. The second term on the right-hand side has no $K(s)$ and the response related to $d(s)$ can be specified by designing $C(s)$, which indicates that Figure 2.21 is a two-degrees-of-freedom control system. It can be seen that they have different structures even though they have the same two-degrees-of-freedom.

2.5.3 Control System with Disturbance Observer and Feedforward

Figure 2.22 shows the block diagram of the control system using both the disturbance observer and the feed-forward term, and the transfer function for the deviation is shown in Equation (2.17):

$$e(s) = \frac{1 - P(s)K(s) + H(PP_0^{-1} - 1)}{1 + P(s)C(s) + H(PP_0^{-1} - 1)}r(s) + \frac{P(s)(1 - H(s))}{1 + P(s)C(s) + H(PP_0^{-1} - 1)}d(s) \tag{2.17}$$

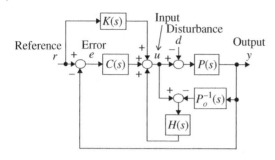

Figure 2.22 DOB and feedforward combined control system.

and if $P(s) = P_0(s)$ holds, then we get Equation (2.18):

$$e(s) = \frac{1 - P(s)K(s)}{1 + P(s)C(s)}r(s) + \frac{P(s)(1 - H(s))}{1 + P(s)C(s)}d(s) \qquad (2.18)$$

This is a two-degrees-of-freedom control system in the sense that the two right-hand-side terms can be designed separately, but $K(s)$ appears only in the reference value response term, $H(s)$ appears only in the term with the disturbance, and $C(s)$ appears in both terms. Such a separation is only possible if $P = P_0$. In reality, there is no ideal match, so the effect of H also appears in the first term on the right-hand side. Assuming that $P = P_0$ still holds, let us check the poles and zeros of the control system.

$P(s) = \frac{N_P(s)}{D_P(s)}$, $H(s) = \frac{N_H(s)}{D_H(s)}$, and $K(s) = \frac{N_K(s)}{D_K(s)}$, and substitute them into Equation (2.17). The feedforward part of the first term on the right-hand side, $K(s)r(s)$, is divided by the second expression on the right-hand side and moved back to give Equation (2.19):

$$e(s) = \frac{1 - \frac{N_P}{D_P}\frac{N_K}{D_K}}{1 + \frac{N_P}{D_P}\frac{N_C}{D_C}}r(s) + \frac{\frac{N_P}{D_P}(1 - \frac{N_H}{D_H})}{1 + \frac{N_P}{D_P}\frac{N_C}{D_C}}d(s)$$

$$= \frac{\left[D_P D_C D_H \quad N_P D_C(D_H - N_H)\right]}{(D_P D_C + N_P N_C)D_H}\begin{bmatrix} r(s) \\ d(s) \end{bmatrix} - \frac{\frac{N_K}{D_K}\frac{N_P}{D_P}r(s)}{1 + \frac{N_P}{D_P}\frac{N_C}{D_C}} \qquad (2.19)$$

The root of the denominator $(D_P D_C + N_P N_C)D_H = 0$ of the first term in the second equation on the right-hand-side of Equation (2.19) is the pole of the control system, which is composed of the pole of the closed-loop system (the root of $D_P D_C + N_P N_C = 0$) and the pole of the disturbance observer (the root of $D_H = 0$).[12] The $D_P D_C D_H$ in the numerator is the numerator of the transfer function with respect to the reference value $r(s)$. It can be observed that the D_H in the denominator and the D_H in the numerator cancel each other out, and

_____

12 The fact that the poles of the control system and the poles of the observer are mutually separated is known as the **separation theorem**, which also holds when the disturbances are included in the state variables, as explicitly stated in Section 5.2.

pole–zero cancelation occurs, eliminating the effect of the disturbance observer on the reference value response.

Meanwhile, the second term of the second equation on the right-hand side is a feedforward input and does not contribute to the stability of the control system.

The roots of $D_P D_C + N_P N_C = 0$, $D_H = 0$, and $D_K = 0$, i.e. the poles of the control system and the poles of the feedforward term, must all have negative stable poles in the real part.

2.6 Effect of Observation Noise and Modeling Error

2.6.1 Effect of Observation Noise

Next, let us examine the effects of **observation noise**. The simulation results for the case where the output noise is turned on at the top of Figure 2.10 are shown in Figure 2.23. Column (a) of Figure 2.23 shows the waveform at $T = 0.01$ s. The estimated waveform in Figure 2.23a-1 for the step-like disturbance is noisy, and if the estimated value is fed back positively, the control input will also be noisy, possibly causing vibration and noise. The noise affects the acceleration and velocity waveforms in Figure 2.23a-2,a-3. Column (b) of Figure 2.23 is the case of $T = 0.5$ s. In the waveform of the disturbance estimate in Figure 2.23b-1, the delay in the estimation increases, but the noise component is almost nonexistent. In the waveforms of Figure 2.23b-2,b-3, the disturbance suppression performance is degraded, and a deviation occurs immediately after $t = 2$ s. Interestingly, the reduction of the noise component in the velocity waveform of Figure 2.23b-3 is small. This is caused by the PI controller. Let us consider the structure of Figure 2.9a again. The control input f is composed of $\bar{f}$ and $\hat{d}$, which are noises in the velocity feedback signal and are amplified by the PI controller. Therefore, there is no choice but to lower the values of the control gains K_p and K_i, which will also lower the response characteristics. Thus, the disturbance observer's design and the controller's design are not independent, and a trade-off is necessary. Eventually, the simplest way to solve the observation noise problem is to use a sensor with high resolution and low noise. If this is impossible, it is necessary to do everything possible, including the disturbance estimation Kalman filter, which will be introduced later.

2.6.2 Effect of Modeling Error

However, if the disturbance $d(t)$ is not due to the external force $f_{ex}(t)$ but due to a **modeling error** of the control target, i.e. internal disturbance, rather than disturbances, we consider the case of internal disturbances. As a concrete example, consider the effects of viscous friction and mass change. That is $P(s) \neq P_o(s)$ in Equation (2.17), which corresponds to the case where the term $H(s)$ appears in the denominator of the control system's characteristic equation.

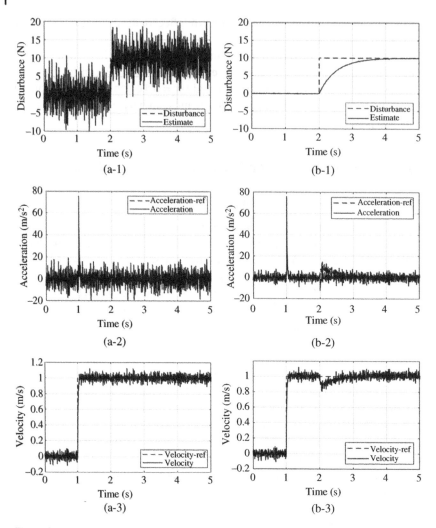

Figure 2.23 Simulation of PI velocity control system (with noise). (a-1) Disturbance waveform at $T = 0.01$ s, (a-2) acceleration waveform at $T = 0.01$ s, (a-3) velocity waveform at $T = 0.01$ s, (b-1) disturbance waveform at $T = 0.5$ s, (b-2) acceleration waveform at $T = 0.5$ s, and (b-3) velocity waveform at $T = 0.5$ s.

2.6.3 Effect of Viscous Friction

Let us examine the influence of viscous friction. Considering that there is viscous friction cv_c that cannot be ignored, and that the disturbance $d(t) = cv_c(t)$, we design a disturbance observer and construct a PI velocity control system with positive feedback of the disturbance estimate $\hat{d}(t)$. Assuming that the mass is nominal

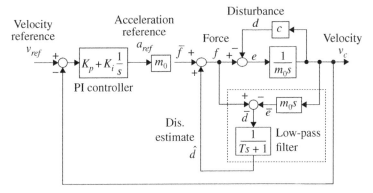

Figure 2.24 PI control system with viscous friction as disturbance.

m_0, Equations (2.20) and (2.21) become

$$m_0\dot{v}_c(t) = f(t) - d(t) \tag{2.20}$$

$$d(t) = cv_c \tag{2.21}$$

Design the PI speed control system shown in Figure 2.24 and draw the **bode diagram**, **pole–zero assignment**, and the unit step response (indicial response).

Figure 2.25 is the result of the simulation. Column (a) on the left is the result of the disturbance observer with a time constant of $T = 0.01$ s, and column (b) on the right is the result with a time constant of $T = 1$ s. The mass $m_0 = 1.0$ kg, the viscous friction coefficient was forced to change to $c = 0.1 \sim 100$ N/(m/s), and the appropriate control gains $K_p = 50$ N/(m/s) and $K_i = 1.0$ N/m were set.

In Figure 2.25a-1, there is no significant change as the coefficient of viscous friction c increases, but in (b-1), the gain decreases. Additionally, Figure 2.25a-2,b-2 show that the poles move. In Figure 2.25a-3, there is no significant change, but in (b-3), the response waveform changes significantly as the viscosity increases. If a small T value is used, the control target can be nominalized by disturbance rejection control (Coulomb friction), but if not, the poles will also move, and it is necessary to recognize that the characteristics of the control system will change.

2.6.4 Effect of Varying Mass

Consider the effects of **mass change**. Suppose that m_0 represents the **nominal value** of mass, which is changed by Δm. In other words, the actual mass is $m = m_0 + \Delta m$. If we assume the cart model to be the mass model for simplicity, the equation of motion becomes Equation (2.22), where

$$(m_0 + \Delta m)\dot{v}_c(t) = f(t) \tag{2.22}$$

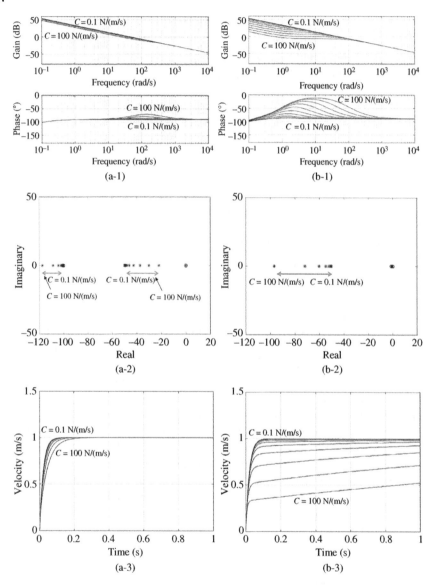

Figure 2.25 Effect of changing the viscous friction on the characteristics. (a-1) Bode diagram of open loop ($T = 0.01$), (a-2) closed-loop pole–zero ($T = 0.01$), (a-3) step response of closed loop ($T = 0.01$), (b-1) bode diagram of open loop ($T = 1$), (b-2) closed-loop pole–zero ($T = 1$), and (b-3) step response of closed loop ($T = 1$).

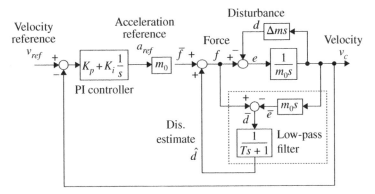

Figure 2.26 PI control system with viscous friction as the disturbance.

The change in the inertia $\Delta m\dot{v}_c$ is treated as a disturbance, and Equations (2.23) and (2.24) considered:

$$m_0\dot{v}_c(t) = f(t) - d(t) \tag{2.23}$$

$$d(t) = \Delta m\dot{v}_c \tag{2.24}$$

Let us design the PI velocity control system shown in Figure 2.26 and draw the **bode diagram**, **pole–zero assignment**, and unit step response (indicial response).

In Figure 2.27, column (a) on the left is performed with the DOB time constant $T = 0.01$ s, and column (b) on the right is performed with the time constant $T = 1$ s. The mass m was extremely varied from 0.1 to 10 times the nominal value of 1 kg, and the same control as in Subsection 2.6.3 was applied.

In Figure 2.25a-1, there is no significant change in the gain curve up to around $1/T$ rad/s even as m increases, while in (b-1), the gain decreases as m increases.

However, the phase changes in Figure 2.27a-1 and b-1 are large. Especially the change in (b-1) is seen in the low frequency. The phase advances when m is small, and when m is large, it is delayed. If m is small, the phase advances, and if m is large, the phase lags. If the phase lags, both the phase and the gain margins decrease. In Figure 2.27a-2,b-2, the poles move significantly, and in (a-2), where the T value is small, the imaginary axis component of the poles becomes more pronounced when the m value becomes large. In Figure 2.27a-3, an oscillation appears, and in (b-3), the delay in the response becomes more pronounced. Thus, it is necessary to pay attention to the magnitude of the effect when controlling by considering the change in inertia force due to mass change as a disturbance.

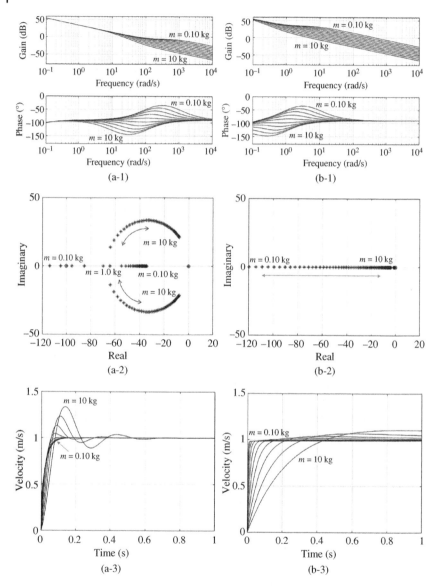

Figure 2.27 Effect of changing the mass on the characteristics. (a-1) Bode diagram of open loop ($T = 0.01$), (a-2) pole–zero of closed loop ($T = 0.01$), (a-3) step response of closed loop ($T = 0.01$), (b-1) bode diagram of open loop ($T = 1$), (b-2) Pole–zero of closed loop ($T = 1$), and (b-3) step response of closed loop ($T = 1$).

2.7 Real System Modeling

Consider implementing a control system using a disturbance observer in a real system. For example, we deal with the case where the **DC motor** is embedded in the mechatronics system and controlled.

2.7.1 DC Motor Torque Control Model

To represent the motion of a DC motor, we basically use the following three equations. Equation (2.25) represents the electrical properties; Equation (2.26) represents the conversion from electrical properties to mechatronics properties; and Equation (2.27) represents the equation of motion of rotation:

$$v_a(t) = R_a i_a(t) + L_a \frac{di_a(t)}{dt} + K_e \omega(t) \tag{2.25}$$

$$\tau(t) = K_t i_a(t) \tag{2.26}$$

$$\tau(t) = J \frac{d\omega(t)}{dt} + \tau_d(t) \tag{2.27}$$

In Equation (2.25), v_a is the drive voltage, i_a is the drive current, ω is the rotation velocity, R_a is the armature resistance, L_a is the armature inductance, and K_e is the induced electromotive force constant. In Equation (2.26), τ is the drive torque, and K_t is the torque constant. J in Equation (2.27) is the moment of inertia of the rotor, ω is the rotation velocity, and τ_d is the disturbance torque. If this is Laplace transformed into a block diagram, it will be on the right side of the voltage v_a of Figure 2.28. In the figure, the area within the single-dotted line can be summarized as $\frac{1}{L_a s + R_a}$. To drive a motor, an inverter circuit such as G_{inv} is used to apply a voltage v_a, and a current i_a generates the torque τ. In other words, since the torque τ cannot be generated directly, even if the disturbance torque τ_d that interferes with the motion can be estimated, it may seem impossible to realize the disturbance suppression control as explained so far.

Therefore, in many cases, a configuration in which a current sensor is used to perform **current feedback** is used, so that the value of the current i_a matches the current reference value i_{ref}. In this case, the current controller $K_i(s)$ is designed

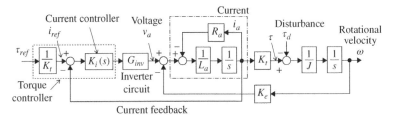

Figure 2.28 Block diagram of the DC motor torque control system.

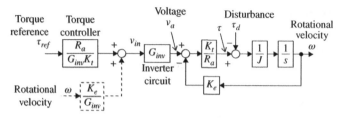

Figure 2.29 DC motor torque control system with voltage control.

to have high gain so that a fast response can be achieved as much as possible. Consequently, $i_a = i_{ref}$ is obtained.

Meanwhile, focusing on Equation (2.26), we can find the relation of $i_{ref} = \frac{1}{K_t}\tau_{ref}$ and insert the gain $1/K_t$ before the electrical current reference value. Consequently, we can achieve $\tau = \tau_{ref}$. This is the part of the **torque controller** in Figure 2.28 that makes it possible to implement most of the control systems introduced in this book.

2.7.2 Without Current Feedback

If the inductance value L_a is small and the electrical time constant $t_e = L_a/R_a$ is significantly smaller than the time constant of the control system, it is acceptable to consider $L_a \doteqdot 0$ and omit the current feedback. In this case, use $1/(L_a + R_a) \doteqdot 1/R_a$ and treat it as shown in Figure 2.29. The dashed line is a compensation function to cancel the induced EMF $K_e\omega$, but it is often omitted if the effect is insignificant.

Note that the case where the current feedback is not performed and the L_a value is not small is treated in Section 4.7.1.[13]

2.7.3 Relationship Between the Cart Model and Rotary-type Motor

Assume that the motor is mounted on a cart and directly connected to a wheel of radius r.[14] The disturbance torque applied to the motor is considered to be the torque τ_c to operate the cart itself, not the torque from external forces. The equation of motion for the motor is Equation (2.28), where the mass of the cart is m_c, the driving force is f_c, and d is the disturbance, and the equation of motion is Equation (2.29):

$$J\dot{\omega}(t) = \tau(t) - \tau_c(t) \tag{2.28}$$

---

13 Some designers may consider the input to be the voltage value v_{in} (or the corresponding PWM value) of Figure 2.29 instead of the torque reference value τ_{ref} without using a torque controller. In that case, we can consider the disturbance not as τ_d but as the equivalent disturbance voltage value $v_{dis} = \frac{R_a}{K_t G_{inv}}\tau_d$ and design an observer to estimate v_{dis}. Here, 'PWM' denotes 'Pulse Width Modulation' for inverter circuits used in motor drive systems.

14 r may also be considered as the feed [m/rad] of a mechatronics system element part such as a ball screw.

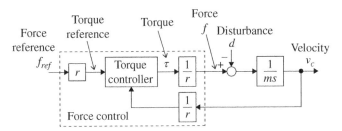

Figure 2.30 Block diagram of the cart model with DC motor.

$$m_c \dot{v}_c(t) = f_c(t) - d(t) \tag{2.29}$$

Here, the relationship between the rotation of the wheel and the velocity of travel is $\omega = v_c/r$, and the relationship between the torque of the motor and the cart's driving force can be expressed as $\tau_c = f_c r$. Equation (2.28) can be rewritten as Equation (2.30):[15]

$$\frac{J}{r} \dot{v}_c = \tau(t) - f_c(r)r \tag{2.30}$$

Dividing both sides of the equation by r and organizing them together with Equation (2.29), we get Equation (2.31) as follows:

$$m\dot{v}_c(t) = f(t) - d(t) \tag{2.31}$$

However, we denote $m = m_c + J/r^2$ and $f(t) = \tau(t)/r$.

Figure 2.30 is the actual block diagram of the mechatronics system with the DC motor in the cart. The dashed part is Figure 2.28, or the dashed line corresponds to τ_{ref} to τ of Figure 2.28 or 2.29, and it corresponds to Equation (2.1) if the torque control without delay is implemented.

2.8   Idea of Robust Control

The term "robust control" means control that is not easily affected by external disturbances or modeling errors due to parameter variations. The following definition is sometimes used:

Definition 2.8.1 (Type of robustness.)   Even in the presence of disturbances and modeling errors:

1. **Robust stability** The control system is stable.
2. **Steady-state robustness** Steady-state error is maintained at 0.
3. **Robust servo system** No deterioration in transient response.

---

15 If the small displacement and small rotation angle can be expressed as $dx_c = rd\theta$, then the virtual work principle $f_c dx_c - \tau d\theta = 0$ is helpful. If we go through $f_c dx_c - \tau d\theta$ and delete $d\theta$ and divide both sides by dt, we get $f_c = \tau_c/r$.

However, there is no rule on whether control according to some design methods can be called robust control or not. Although some may disagree, let us consider it along with the history of control theory.

It is clear from the industry's long history of control theory that **classical control theory**, which assumes a SISO system, is still the most practical theory since it was systematized in the 1950s. Since the 1960s, **modern control theory** has been proposed to use equations of state that can handle MIMO systems, and the concepts of **controllable, observable, optimal control**, and observer were devised. The value of these concepts and methods, which did not exist in classical control, is great, but they are far from being applicable in the frequency domain; highly accurate models are needed. Although created as an application of modern control theory, the disturbance observer is considered revolutionary in that it can handle uncertainties and unknown disturbances. Later, since the 1980s, research on **robust control theory** has been emphasized, and control theory using coprime factorization has been developed. After that, H_∞ control and μ design methods have been proposed [8–11]. The sliding mode control [12] also attracted attention simultaneously, and a design example combined with a disturbance observer was proposed [13, 14].

Generally, the control theories related to robustness are collectively referred to as "robust control theories." For example, H_∞ control was groundbreaking because it defined additive and multiplicative uncertainties and used the **small gain theorem** to design a controller that guarantees **internal stability** under uncertainty. However, the resulting control systems did not necessarily take root in the industrial world, probably because there were many cases where only conservative control performance could be obtained. Adaptive control theory, which assumes that even the parameter values are unknown or changeable, is supposed to be the ultimate in robust control, but even small fluctuations can destabilize it. The establishment of simple adaptive control [15, 16] is the subject of the ongoing research. In the world of control engineering research, it is often debated which style is superior or inferior, but each style has its advantages and disadvantages, and they all seem to be evolving gradually through the efforts of researchers.

In contrast to this history, what do we mean when we refer to a control system with a disturbance observer as "robust control"? If we can implement a disturbance observer with a small value of the time constant T and positive feedback of the disturbance estimate, we can confirm that the effect of disturbance rejection is significant in many cases, and that there is little change in control performance and stability in response to parameter changes, so we can say that we have achieved a robust control system.' The disturbance suppression performance in the frequency domain can also be shown quantitatively by setting the time constant T of the low-pass filter. In many examples using transfer function expressions, the robustness is often discussed or evaluated in the framework of classical control theory, as in this chapter.

In contrast, the disturbance observer and the combined control system in the state-space representation introduced in Chapter 4 and later are designed using the general **pole assignment method** or the **LQ control method**. However, there are not many examples showing considerations in line with robust control theory, such as guaranteeing stability by quantitatively defining in advance the maximum value of disturbance and the range of parameter variation, but some papers have reported that it can be interpreted in relation to the stabilizing compensator in the **coprime factorization** representation, the free parameter design method, and the *H∞* control [11, 17–22]. In this book, we also introduce the relationship between the coprime factorization and the disturbance observer in Chapter 3.

It seems to me that the "disturbance observer" and the "control system with disturbance observer" are most valuable because they focus on a physical quantity called "disturbance" and are estimation and control methods that are closely related to the plant. We also believe that theories and technologies can be integrated and developed with various control theories and methods in the future, as introduced in this book.

References

1 Toshiaki Murakami, Fangming Yu, Kouhei Ohnishi: Torque sensorless control in multidegree-of-freedom manipulator, IEEE Transactions Industrial Electronics,Vol. 40, No. 2, 259–265, 1993.

2 Makoto Iwasaki, Tomohiro Shibata, Nobuyuki Matsui: Disturbance-observer-based nonlinear friction compensation in table drive system, IEEE/ASME Transactions on Mechatronics, Vol. 4, No. 1, 3–8, 1999.

3 Ryo Kikuuwe, Naoyuki Takesue, Akihito Sano, Hiromi Mochiyama, Hideo Fujimoto: Admittance and impedance representations of friction based on implicit Euler integration, IEEE Transactions on Industrial Electronics, Vol. 22, No. 6, 1176–1188, 2006.

4 Satoshi Komada, Kouhei Ohnishi: Force feedback of robot manipulator by the acceleration tracking orientation method, IEEE Transactions Industrial Electronics,Vol. 37, No. 1, 6–12, 1990.

5 Kouhei Ohnishi, Masaaki Shibata, Toshiyuki Murakami: Motion control for advanced mechatronics, IEEE/ASME Transactions on Mechatronics, Vol. 1, No. 1, 56–67, 1996.

6 Naoki Motoi, Tomoyuki Shimono, Ryogo Kubo, Atsuo Kawamura: Force-based Variable Compliance Control Method for Bilateral System with Different

Degrees of Freedom, The 12th IEEE International Workshop on Advanced Motion Control (AMC'12), 1–6, 2012.

7 Anh-Van Ho, Shinichi Hirai: Mechanics of Localized Slippage in Tactile Sensing: And Application to Soft Sensing Systems, Springer, 2014.

8 John C. Doyle, Keith Glover, Pramod P. Khargoneker, Bruce A. Frncis: State-space solutions to standard H_2 and H_∞ control problems, IEEE Transactions on Automatic Control, Vol. 34, No. 8, 831–847, 1989.

9 Bruce A. Francis: A Cource in H_∞ Control Theory, Lecture Notes in Control and Information Sciences, Springer-Verlag, 1987.

10 Kemin Zhou, John C. Doyle, Keith Glover: Robust and Optimal Control, Prentice Hall, 1995.

11 Tstomu Mita, Mitsuo Hirata, Kenichi Murata, Hui Zhang: H_∞ control versus disturbance-observer-based control, IEEE Transactions on Industrial Electronics, Vol. 45, No. 3, 488–495, 1998.

12 Vadim Utkin: Sliding Modes in Control and Optimization, Springer-Verlag, 1992.

13 Yasutaka Fujimoto, Atsuo Kawamura: Robust servo-system based on two-degree-of-freedom control with sliding mode, IEEE Transactions on Industrial Electronics, Vol. 42, No. 3, 272–280, 1995.

14 Akira Shimada, Kenzo Nonami, Isamu Kai, Youngwei Shi, Makoto Ijiri, Seiichi Suzuki: Variable Capacity Control of Compressor for Automotive Air-Conditioning System using Reference Model Following Based Sliding Mode Control, The 6th International Conference on Motion and Vibration Control, 306–311, 2002.

15 Petros Ioannou, Jing Sun: Robust Adaptive Control, Dover Publications, 2012.

16 Itzhak Barkana: Simple adaptive control - a stable direct model reference adaptive control methodology - brief survey, International Journal of Adaptive Control and Signal Processing, Vol. 28, No. 7–8, 567–603, 2014.

17 Tomoharu Doi, Koichi Osuka, Toshiro Ono, Ryoji Kwatani: A study on the robust stability for disturbance observers, Transactions of SICE, Vol. 34, No. 10, 1404–1410, 1998 (In Japanese).

18 Shiro Masuda, Valeri T. Kroumov, Akira Inoue, Kenji Sugimoto: A Design Method of Multivariable Model Reference Adaptive Control System Using Coprime Factorization Approach, Proceedings of the 32nd Automatic Control Conference on Decision and control, FM4-3:10, 1993.

19 Kazunori Matsumoto Tatsuya Suzuki Shigeru Okuma: The design method of two-degree-of-freedom controller using μ-synthesis and its application to two-mass system, IEEE 4th International Workshop on Advanced Motion Control (AMC'96), 606–611, 1996.

20 Akitaka Mizutani, Kazuhiro Yubai, Junji Hirai: A Direct Design from Input/Output Data of the Youla Parameter for Compensating Plant Perturbation on GIMC Structure, IEEE 35th Annual Conference of IEEE Industrial Electronics, 3047–3052, 2009.

21 Kiyoshi Ohishi, Toshimasa Miyazaki, Yoshihiro Nakamura: High performance ultra-low speed servo system based on doubly coprime factorization and instantaneous speed observer, IEEE Transactions on Mechatronics, Vol. 1, No. 1, 89–98, 1996.

22 Chun-Chih Wang, Masayoshi Tomizuka: Design of Robustly Stable Disturbance Observers Based on Closed Loop Consideration Using H∞ Optimization and its Applications to Motion Control Systems, Proceedings of 2004 American Control Conference, ThP16.1, 3764–3769, 2004.

3

Stabilized Control and Coprime Factorization

Modeling methods and control system design methods based on coprime factorization have been proposed as the basic theory of robust control theory. This chapter introduces an overview of coprime factorization, followed by an explanation of its relationship with disturbance observers and two-degrees-of-freedom control.

3.1 Coprime Factorization and Derivation of Stabilizing Controller

When the control plant $P(s)$ is proper, the $P(s)$ can be decomposed $P(s)$ into $N(s)D - 1(s) = \tilde{D} - 1(s)\tilde{N}(s)$. Where we call $N(s), D(s) \in RH_\infty$ right coprime, and $\tilde{N}(s), \tilde{D}(s) \in RH_\infty$ left coprime $\tilde{N}(s), \tilde{D}(s) \in RH_\infty$ and can obtain the following equation decomposed into $P(s) = N(s)D^{-1}(s) = \tilde{D}^{-1(s)}\tilde{N}(s)$.[1]

The former is the right coprime factorization form, and the latter is the left coprime factorization form. Additionally, a controller that is $u(s) = -C(s)y(s)$ with the observer and the state feedback together is expressed as $C(s) = \tilde{X}^{-1}(s)\tilde{Y}(s) = Y(s)X^{-1}(s)$ using $X, Y, \tilde{X}, \tilde{Y} \in RH_\infty$.[2] These parameter representations are called Yula parameterization. This chapter shows the relationship between the control system [2, 3] and the disturbance observer [4, 5] using these parameters.

1 If the "degree of the denominator polynomial $\geq$ that of the numerator polynomial" of the transfer function, then we say that the transfer function is proper. Furthermore, if $>$, we say it is strictly proper. Then, we express the entire rational function that is proper and stable by RH_∞. RH_∞ also includes constants. Also, right (or left) coprime means the absence of unstable common factors at the zeros. For more details, see [1], for example.
2 In this book, we define $A_F = A - BF$ and $A_H = A - HC$, but many references define $A_F + BF$ and $A_H = A + HC$. Since the definitions of $X, \tilde{X}$, etc. vary, the signs and order in the **coprime factorization** and stabilization controller introduced below also vary based on the literature.

Disturbance Observer for Advanced Motion Control with MATLAB/Simulink, First Edition. Akira Shimada.
© 2023 The Institute of Electrical and Electronics Engineers, Inc. Published 2023 by John Wiley & Sons, Inc.
Companion website: www.wiley.com/go/disturbanceobserver

3.1.1 Derivation of Parameters for Coprime Factorization

Derivation of $P(s) = N(s)D^{-1}(s)$

The state equation, output equation, and state feedback of the control plant $P(s)$ are defined as follows:

$$\dot{x} = Ax + Bu, y = Cx \tag{3.1}$$

$$u = -Fx + v \tag{3.2}$$

The v is a superficial input used for convenience in the derivation. Combining both equations and setting $A_F = A - BF$, we obtain Equation (3.3):

$$\dot{x} = (A - BF)x + Bv = A_F x + Bv \tag{3.3}$$

Making Equation (3.2) think like an output equation with the input as v and the output as u, and using $\hat{x}(s) = (sI - A_F)^{-1}Bv(s)$ derived from Equation (3.3), we can get Equation (3.4):[3]

$$D(s) = -F(sI - A_F)^{-1}B + I = [A_F, B, -F, I] \tag{3.4}$$

A_F is stable due to state feedback, and $D(s)$ is stable and proper because of I. $D(s)$ has a proper D^{-1}. Since $u(s) = D(s)v(s)$, we have $v(s) = D^{-1}(s)u(s)$, and from Equations (3.3) and (3.1), we get Equation (3.5):

$$N(s) = C(sI - A_F)^{-1}B = [A_F, B, C, 0] \tag{3.5}$$

Since $y(s) = N(s)v(s)$ holds, using $v(s) = D^{-1}(s)u(s)$, v vanishes, and it leads to $y(s) = N(s)D^{-1}(s)u(s)$. In other words, $P(s) = N(s)D^{-1}(s)$.

Derivation of $P(s) = \tilde{D}^{-1}(s)\tilde{N}(s)$

Derive the dual $P(s) = \tilde{D}(s)\tilde{N}(s)$ using the identity observer. As $A_H = A - HC$,

$$\dot{\hat{x}} = A\hat{x} + Bu + H(y - \hat{y}) = A_H\hat{x} + Bu + Hy \tag{3.6}$$

Performing the Laplace transform and organizing it with $\hat{y} = C\hat{x}$, we get Equation (3.7):

$$\hat{y}(s) = C\hat{x}(s) = C(sI - A_H)^{-1}Bu(s) + C(sI - A_H)^{-1}Hy(s) \tag{3.7}$$

Define Equation (3.8) and replace it as $\tilde{y}(s) = \tilde{N}(s)u(s)$ to get Equation (3.9):

$$\tilde{N}(s) = C(sI - A_H)^{-1}B = [A_H, B, C, 0] \tag{3.8}$$

$$\tilde{y}(s) = -C(sI - A_H)^{-1}Hy(s) + I \cdot \hat{y}(s) \tag{3.9}$$

---

3 Appendix A.4 contains a supplementary explanation of Doyle's notation.

From the output estimation error $e = y - \hat{y}$, $\hat{y}(s) = y(s) - e(s)$. Assuming that this output estimation deviation $e(t) \to 0$, replace $\hat{y}(s)$ with $y(s)$ to obtain Equation (3.10):

$$\tilde{y}(s) = -C(sI - A_H)^{-1}Hy(s) + I \cdot y(s) = \tilde{D}(s)y(s) \tag{3.10}$$

However, we defined $\tilde{D}$ as Equation (3.11):

$$\tilde{D}(s) = -C(sI - A_H)^{-1}H + I = [A_H, H, -C, I] \tag{3.11}$$

Consequently, $\tilde{y}(s) = \tilde{D}(s)y(s)$, which leads to Equation (3.12):

$$y(s) = \tilde{D}^{-1}(s)\tilde{y}(s) = \tilde{D}^{-1}(s)\tilde{N}(s)u(s) \tag{3.12}$$

In other words, we get $P(s) = \tilde{D}^{-1}(s)\tilde{N}(s)^4$.

Derivation of $C(s) = \tilde{X}^{-1}(s)\tilde{Y}(s)$

Assume a control system using an observer, combing Equation (3.6) and $u = -F\tilde{x}$, we derive $u = -F(sI - A_H)^{-1}Bu - F(sI - A_H)^{-1}Hy$, and next, we can obtain Equation (3.13)

$$u(s) = -\{I + F(sI - A_H)^{-1}B\}^{-1}\{F(sI - A_H)^{-1}H\}y(s) \tag{3.13}$$

where $\tilde{X}$ and $\tilde{Y}$ are defined as in Equations (3.14) and (3.15):

$$\tilde{X}(s) = I + F(sI - A_H)^{-1}B = [A_H, B, F, I] \tag{3.14}$$

$$\tilde{Y}(s) = F(sI - A_H)^{-1}H = [A_H, H, F, 0] \tag{3.15}$$

Then, we have $u(s) = -\tilde{X}^{-1}(s)\tilde{Y}(s)y(s)$, and the controller is represented as $C(s) = \tilde{X}^{-1}(s)\tilde{Y}(s)$.

Derivation of $C(s) = Y(s)X^{-1}(s)$

Derive a controller that is dual to the above. The basic equation for the following identity observer is

$$\dot{\hat{x}} = A\hat{x} + Bu + H(y - \hat{y}) = A_F\hat{x} + He$$

Using $u = -F\hat{x}$ and $\hat{y} = C\hat{x}$, we get Equation (3.16):

$$u(s) = -F\hat{x}(s) = -F(sI - A_F)^{-1}He(s) \tag{3.16}$$

Next, defining Y as in Equation (3.17), we have $u = -Y(s)e(s)$:

$$Y(s) = F(sI - A_F)^{-1}H = [A_F, H, F, 0] \tag{3.17}$$

---

4 Although state feedback and observer design are not the only ways to perform coprime factorization, only this design process is shown in this document.

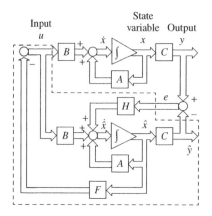

Input
State
variable Output
u
$\dot{x}$ x   y

Figure 3.1 Block diagram of the combined observer control system to illustrate the dashed section.

We only need to be able to express the remaining $e(s) = X^{-1}(s)y(s)$. In Figure 3.1, the part from e to $\hat{y}$ in the dashed frame is represented by the following transfer function:

$$\hat{y}(s) = \{-C(sI - A_F)^{-1}H\}e(s)$$

However, we obtain the following equation by considering $y = \hat{y} + e$ from $e = y - \hat{y}$:

$$y(s) = \{-C(sI - A_F)^{-1}\}He(s) + e(s) = \{C(sI - A_F)^{-1}H + I\}e(s)$$

If we set $X(s) = C(sI - A_F)^{-1}H + I$, then after $y(s) = X(s)e(s)$, we get $e(s) = X^{-1}(s)y(s)$. From this, we get Equation (3.18):

$$u(s) = -Y(s)X^{-1}(s)y(s) \tag{3.18}$$

Theorem 3.1.1 *When we have right and left coprime factorizations of $P(s)$, creating $X(s), Y(s), \tilde{X}(s), \tilde{Y}(s)$ in RH_∞,* **double coprime factorization** *is expressed as Equation (3.19):*

$$\begin{bmatrix} \tilde{X}(s) & \tilde{Y}(s) \\ \tilde{N}(s) & -\tilde{D}(s) \end{bmatrix} \begin{bmatrix} D(s) & Y(s) \\ N(s) & -X(s) \end{bmatrix}$$

$$= \begin{bmatrix} \tilde{X}D + \tilde{Y}N & \tilde{X}Y - \tilde{Y}X \\ \tilde{N}D - \tilde{D}N & \tilde{N}Y + \tilde{D}X \end{bmatrix} = \begin{bmatrix} I & 0 \\ 0 & I \end{bmatrix} = I \tag{3.19}$$

For the right coprime factorization $P = ND^{-1}, \tilde{X}D + \tilde{Y}N = I$ holds, and $P = \tilde{D}^{-1}\tilde{N}$, $\tilde{N}Y + \tilde{D}X = I$ holds for the left coprime factorization. These are called **Bezu equality**.[5]

---

5 The notation of the double coprime factorization is not unique. It depends on how the variables are defined and signed. The checking process is shown in Appendix A.4.

Example 3.1.1 (Example of a single-input single-output system.) Assume a spring-mass damper system (mass $m = 1$ kg, viscosity $c = 3$ N/(m/s), and elasticity $k = 2$ N/m). Then, we can design a state feedback control system with poles at -2 and -3 and with poles of the observer at -3 and -5. We can create A, B, C, F, and H, and it yields $P(s) = C(sI - A)^{-1}B$, where $P(s) = \frac{1}{(s+1)(s+2)}$. With these as the materials, we can calculate each parameter:

$$N(s) = \frac{1}{(s+2)(s+3)}, \quad D(s) = \frac{(s+1)(s+2)}{(s+2)(s+3)}$$

$$P(s) = N(s)D^{-1}(s) = \frac{1}{(s+2)(s+3)}\frac{(s+2)(s+3)}{(s+1)(s+2)} = \frac{1}{(s+1)(s+2)}$$

$$\tilde{N}(s) = \frac{1}{(s+3)(s+5)}, \quad \tilde{D}(s) = \frac{(s+1)(s+2)}{(s+3)(s+5)}$$

$$P(s) = \tilde{D}^{-1}(s)\tilde{N}(s) = \frac{(s+3)(s+5)}{(s+1)(s+2)}\frac{1}{(s+3)(s+5)} = \frac{1}{(s+1)(s+2)}$$

$$X(s) = \frac{s^2 + 10s + 29}{(s+2)(s+3)} = \frac{(s+5+2j)(s+5-2j)}{(s+2)(s+3)}$$

$$Y(s) = \frac{16(s+2)}{(s+2)(s+3)}$$

$$C(s) = Y(s)X^{-1}(s) = \frac{16(s+2)}{s^2 + 10s + 29} = \frac{16(s+2)}{(s+5+2j)(s+5-2j)}$$

$$\tilde{X}(s) = \frac{s^2 + 10s + 29}{(s+3)(s+5)} = \frac{(s+5+2j)(s+5-2j)}{(s+3)(s+5)}$$

$$\tilde{Y}(s) = \frac{16(s+2)}{(s+3)(s+5)}$$

$$C(s) = \tilde{X}^{-1}(s)\tilde{Y}(s) = \frac{16(s+2)(s+3)(s+5)}{(s^2 + 10s + 29)(s+3)(s+5)}$$

$$= \frac{16(s+2)}{(s+5+2j)(s+5-2j)}$$

$$\tilde{X}D + \tilde{Y}N = \frac{(s^2 + 10s + 29)}{(s+3)(s+5)}\frac{(s+1)(s+2)}{(s+2)(s+3)} + \frac{16(s+2)}{(s+3)(s+5)}\frac{1}{(s+2)(s+3)}$$

$$= \frac{(s^2 + 10s + 29)(s+1)(s+2) + 16(s+2)}{(s+3)^2(s+5)(s+2)} = 1$$

$$\tilde{N}Y + \tilde{D}X = \frac{1}{(s+3)(s+5)}\frac{16(s+2)}{(s+2)(s+3)} + \frac{(s+1)(s+2)}{(s+3)(s+5)}\frac{(s^2 + 10s + 29)}{(s+2)(s+3)}$$

$$= \frac{16(s+2) + (s+1)(s+2)(s^2 + 10s + 29)}{(s+3)(s+5)(s+2)(s+3)} = 1$$

The $(s+2)$ and $(s+3)$ corresponding to the poles of the control system and $(s+3)$ and $(s+5)$ corresponding to the poles of the observer appear in the denominator and numerator. These indicate the occurrence of pole-zero cancelation.

Each of the above variables can be calculated by a simple sum-of-products operation since they are expressed as the transfer functions of a single-input single-output (SISO) system. However, in the case of a state-space representation, they can be derived using Doyle notation, as shown in Appendix A.4.

3.1.2 Stabilizing Controller and Free Parameters

The following process explains a stabilizing controller that uses a control system with state feedback and an observer and incorporates free parameters $Q(s)$ [1, 2]. From Equation (3.7),

$$\hat{y}(s) = C\hat{x}(s) = C(sI - A_H)^{-1}Hy(s) + C(sI - A_H)^{-1}Bu(s)$$
$$= \{-\tilde{D}(s) + I\}y(s) + \tilde{N}(s)u(s) \tag{3.20}$$

and express the output deviation $e(s)$ using Equation (3.20) as in Equation (3.21),

$$e(s) = y(s) - \hat{y}(s) = \tilde{D}(s)y(s) - \tilde{N}(s)u(s) \tag{3.21}$$

Let us try to find the control input with $-Q(s)e(s)$ added to the state feedback.[6]

$$u(s) = -F\hat{x}(s) - Q(s)e(s)$$
$$= -F(sI - A_H)^{-1}Hy - F(sI - A_H)^{-1}Bu(s) - Q(s)e(s)$$
$$= -\tilde{Y}(s)y(s) + \{I - \tilde{X}(s)\}u(s) + Q(s)\{\tilde{N}(s)u(s) - \tilde{D}(s)y(s)\} \tag{3.22}$$

Organizing Equation (3.22), we get Equation (3.23):

$$u(s) = -\{\tilde{X}(s) - Q(s)\tilde{N}(s)\}^{-1}\{\tilde{Y}(s) + Q(s)\tilde{D}(s)y(s)\}$$

Theorem 3.1.2 *Using the double coprime factorization, all stabilizing controllers $C(s)$ for the control plant $P(s)$ are represented by Equation (3.23) with free parameter $Q(s) \in RH_\infty$:*

$$C(s) = \{\tilde{X}(s) - Q(s)\tilde{N}(s)\}^{-1}\{\tilde{Y}(s) + Q(s)\tilde{D}(s)\} \tag{3.23}$$

$$= \{Y(s) + D(s)Q(s)\}\{X(s) - N(s)Q(s)\}^{-1} \tag{3.24}$$

However, $\det(\tilde{X} - Q\tilde{N}) \neq 0$ and $\det(X - NQ) \neq 0$, which are automatically satisfied if $P(s)$ is strictly proper.

---

6 We are not sure whether to use $+Q(s)e(s)$ or $-Q(s)e(s)$, but we chose $-d(s)$ as the sign to cancel.

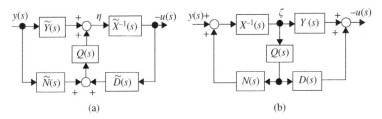

Figure 3.2 Block diagram of stabilization controller $C(s)$. (a) Block diagram of Equation (3.23) and (b) block diagram of Equation (3.24).

$C(s)$ can be expressed as a transfer function block diagram as in Figure 3.2. Using $\eta = \tilde{Y}y + Q\tilde{N}y + Q\tilde{D}(-u)$ from Figure 3.2a, we find $-u = \tilde{X}^{-1}\eta = \tilde{X}^{-1}(\tilde{Y}y + Q\tilde{N}y + Q\tilde{D})(-u)$ and organize it. It follows that $-(I - \tilde{X}^{-1}Q\tilde{N})u = \tilde{X}^{-1}(\tilde{Y} + Q\tilde{D})y$, and multiplying by $\tilde{X}$ from the left and solving for $-u$, we obtain Equation (3.23).

For Figure 3.2b, Equation (3.24) can be summarized using $-u = Y\zeta$. The advantage of using the Yula parameter is that using Equation (3.23) (or Equation (3.24)) as the controller guarantees the following **internal stability**, and the free parameter Q(s) appears in the feedback system's linear transfer function.

> **Theorem 3.1.3** *A closed-loop system of Figure 3.3 is said to be internally stable if the propriety G(s) and H(s) of the closed-loop system have no unstable pole-zero cancelation and the relation between certain inputs and outputs is stable.*

For example, in the case of a single-input single-output system, when u and n are the inputs and y and w are the outputs of Figure 3.3, the input–output relationship is expressed by the following equation, where the four transfer functions on the right-hand side are proper and stable, a necessary and sufficient condition for internal stability:

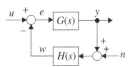

Figure 3.3 Block diagram of a closed loop system.

$$\begin{bmatrix} y(s) \\ w(s) \end{bmatrix} = \begin{bmatrix} \dfrac{G(s)}{1 + G(s)H(s)} & -\dfrac{G(s)H(s)}{1 + G(s)H(s)} \\ \dfrac{G(s)H(s)}{1 + G(s)H(s)} & \dfrac{(s)H(s)}{1 + G(s)H(s)} \end{bmatrix} \begin{bmatrix} u(s) \\ n(s) \end{bmatrix}$$

3.1.3 Double Coprime Factorization Involving $Q(s)$

For Equation (3.19), multiply the matrix containing $Q(s)$ from left to right.

$$\begin{bmatrix} I & -Q(s) \\ 0 & I \end{bmatrix} \begin{bmatrix} \tilde{X}(s) & \tilde{Y}(s) \\ \tilde{N}(s) & -\tilde{D}(s) \end{bmatrix} \begin{bmatrix} D(s) & Y(s) \\ N(s) & -X(s) \end{bmatrix} \begin{bmatrix} I & Q(s) \\ 0 & I \end{bmatrix}$$

$$= \begin{bmatrix} I & -Q(s) \\ 0 & I \end{bmatrix} \begin{bmatrix} I & 0 \\ 0 & I \end{bmatrix} \begin{bmatrix} I & Q(s) \\ 0 & I \end{bmatrix} = \begin{bmatrix} I & 0 \\ 0 & I \end{bmatrix}$$

If we list them component by component, we get four Equations from (3.25) to (3.28).

$$(\tilde{X} - Q\tilde{N})D + (\tilde{Y} + Q\tilde{D})N = I \tag{3.25}$$

$$(\tilde{X} - Q\tilde{N})(\tilde{Y} + Q\tilde{D}) = (\tilde{Y} + Q\tilde{D})(\tilde{X} + NQ) \tag{3.26}$$

$$\tilde{N}D = \tilde{D}N \tag{3.27}$$

$$\tilde{N}(Y + DQ) + \tilde{D}(X - NQ)N = I \tag{3.28}$$

Equations (3.25) and (3.28) are called Bezout's identity respectively, and Equations (3.23) and (3.24) are obtained when we organize Equations (3.26) and (3.27).[7]

3.2 Relationship with Disturbance Observer

Organizing Equation (3.22) into Equation (3.29), and multiplying both sides by $\tilde{X}^{-1}(s)$ from the left, we get Equation (3.30).

$$\tilde{X}(s)u(s) = -\tilde{Y}(s) + Q(s)\{\tilde{N}(s)u(s) - \tilde{D}(s)y(s)\} \tag{3.29}$$

$$u(s) = -\tilde{X}^{-1}(s)\tilde{Y}(s)y(s) + \tilde{X}^{-1}Q(s)\{\tilde{N}(s)u(s) - \tilde{D}(s)y(s))\} \tag{3.30}$$

This is illustrated in Figure 3.4.

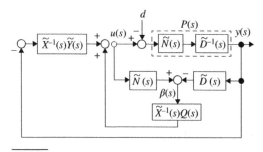

Figure 3.4 Block diagram of a stabilized control system with free parameters.

---

7 If we swap the matrices containing $Q(s)$ before and after, the result will still be a unitary matrix, but the () in Equations (3.25)–(3.28) will have the sign reversed.

Observing this structure, it appears that the disturbances are estimated by the disturbance observer, suppressed by the positive feedback, and stabilized by the controller $\tilde{X}^{-1}(s)\tilde{Y}(s)$. Now, let us find out what $\beta(s)$ in the figure represents: $\beta(s) = \tilde{N}(s)u(s) - \tilde{D}(s)y(s) = \tilde{N}(s)u(s) - \tilde{D}(s)\{\tilde{D}^{-1}(s)\tilde{N}(s)\}\{u(s) - d(s)\} = \tilde{N}(s)d(s)$. For a stepwise disturbance $d(s) = d_0/s$, we can use the **final value theorem** to get $\beta(\infty) = \lim_{s \to 0} s\tilde{N}(s)\frac{d_0}{s} = \tilde{N}(0)$. If $\tilde{N}(0) \neq 0$ from d_0, then d_0 can be calculated [4].

3.3 Coprime Factorization and Structure of Two-degrees-of-freedom Control System

Regarding the control system using coprime factorization, consider the two-degrees-of-freedom control system. Using $N(s)K(s)$ as the feedback control system and $D(s)K(s)$ as the feedforward input, we obtain Figure 3.5.

How should $Q(s)$ be designed? The most straightforward design is to eliminate the disturbance, corresponding to Chapter 2.

$$\hat{d}(s) = \tilde{X}^{-1}(s)Q(s)\beta(s) = \tilde{X}^{-1}Q(s)\tilde{N}d(s) = H(s)d(s)$$

It is better to set $Q(s) = \tilde{X}(s)H(s)\tilde{N}^{-1}(s)$, so that $Q(s) = \tilde{X}(s)H(s)\tilde{N}^{-1}(s)$, but since $\tilde{N}^{-1}$ is not proper, the low-pass filter $F(s) = \frac{\omega^2}{s^2+2\zeta\omega s+\omega^2}$ and multiply it by $Q(s) = \tilde{X}(s)H(s)\tilde{N}^{-1}(s)F(s)$. Here, if the zeros of $\tilde{N}(s)$ are unstable, the poles of $\tilde{N}^{-1}(s)$ become unstable, and $Q(s) \in RH_\infty]$ does not become $Q(s) \in RH_\infty$.[8] An example program is shown in List 3.1, and the Simulink model and simulation examples are shown in Figures 3.6 and 3.7.

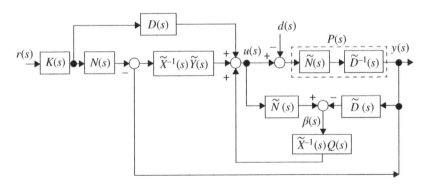

Figure 3.5 Block diagram of a two-degrees-of-freedom control system including $\tilde{Q}(s)$.

---

8 $H(s)$ is the $H(s)$ introduced in Chapter sec:basic. Instead of adding $F(s)$, we can regard $H(s)$ as being designed considering the relative order of different numbers.

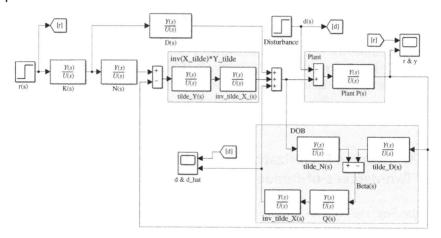

Figure 3.6 Example of a Simulink model of a two-degrees-of-freedom control system.

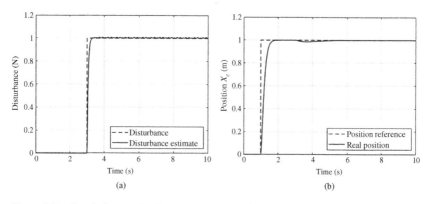

Figure 3.7 Simulation results for a two-degrees-of-freedom control system. (a) Disturbance waveform and (b) position waveform.

List 3.1: An example of a two-degrees-of-freedom control system using coprime factorization.

```
1   %% Physical parameters and mathematical models
2   % for spring−mass−damper systems
3   m=1;c=3;k=2;A=[0,1;−k/m,−c/m];B=[0;1/m];C=[1,0];
4   [num,den]=ss2tf(A,B,C,0);P=tf(num,den);[z,p,k]=tf2zp(num,den);
5
6   %% Control System Design and Observer Design
7   pole=[−2,−3];F=place(A,B,pole);Af=A−B*F;
8   pole_ob=[−3,−5];H_temp=place(A',C',pole_ob);H=H_temp';Ah=A−H*C;
9
```

```
10   %% P_nd=N*inv(D)
11   [num_D,den_D]=ss2tf(Af,B,-F,1);D=tf(num_D,den_D);
12   [num_N,den_N]=ss2tf(Af,B,C,0);N=tf(num_N,den_N);P_nd=N*inv(D);
13   [n_nd,d_nd]=tfdata(P_nd,'v');[z_nd,p_nd,k_nd]=tf2zp(n_nd,d_nd);
14
15   %% tild_P=inv(tilde_D)*tilde_N
16   [num_tN,den_tN]=ss2tf(Ah,B,C,0);tilde_N=tf(num_tN,den_tN);
17   [num_tD,den_tD]=ss2tf(Ah,H,-C,1);tilde_D=tf(num_tD,den_tD);
18   P_tilde=inv(tilde_D)*tilde_N;[n_til,d_til]=tfdata(P_tilde,'v');
19   [z_til,p_til,k_til]=tf2zp(n_til,d_til);
20
21   %% Checking controllers and double coprime factorization
22   % Creation of X, Y
23   [num_X,den_X]=ss2tf(Af,H,C,1);X=tf(num_X,den_X);[z_X,p_X,k_X]=tf2zp(num_X,den_X);
24   [num_Y,den_Y]=ss2tf(Af,H,F,0);Y=tf(num_Y,den_Y);[z_Y,p_Y,k_Y]=tf2zp(num_Y,den_Y);
25   %
26   C_XY=Y*inv(X);[num_XY,den_XY]=tfdata(C_XY,'v');
27   [z_C_XY,p_C_XY,k_C_XY]=tf2zp(num_XY,den_XY);
28
29   %% Creation of tilde_X, tilde_Y
30   [num_til_X,den_til_X]=ss2tf(Ah,B,F,1);tilde_X=tf(num_til_X,den_til_X);
31   [z_til_X,p_til_X,k_til_X]=tf2zp(num_til_X,den_til_X);
32   [num_til_Y,den_til_Y]=ss2tf(Ah,H,F,0);tilde_Y=tf(num_til_Y,den_til_Y);
33   [z_til_Y,p_til_Y,k_til_Y]=tf2zp(num_til_Y,den_til_Y);
34
35   %% Design of Q(s) and K(s)
36   T=0.05;num_H=1;den_H=[T,1];H=tf(num_H,den_H);
37   num_Q=conv(conv(num_til_X,num_H),den_tN);
38   num_tN_2=conv(num_tN,[0.01,1]);
39   den_Q=conv(conv(den_til_X,den_H),num_tN_2);
40   %
41   zeta=1;w=10; % Parameter for making K(s) proper
42   num_K=den_N*w^2;den_K=conv(num_N,[1,2*zeta*w,w^2]);
43
44   %% Simulation
45   open_system('sim_Figure_3_6_and_3_7_yula_Qcon');
46   z=sim('sim_Figure_3_6_and_3_7_yula_Qcon');
```

In Figure 3.6, the numerator and denominator of the transfer function are set corresponding to the variable names noted at the bottom of each transfer function block. Additionally, the simulation results of Figure 3.7 show that a stable control system can be realized, and disturbance estimation is well performed.[9]

---

9 For tf, tfdata, tf2zp, ss2tf, conv, etc. in the program list, please refer to the MathWorks company's website.

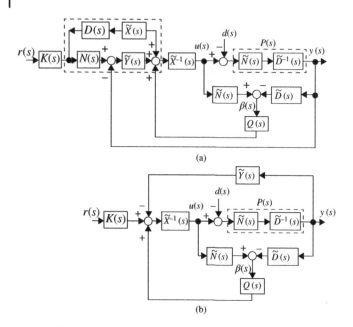

(a)

(b)

Figure 3.8 Block diagram of a two-degrees-of-freedom control system including $Q(s)$. (a) Block diagram with double coprime factorization in mind and (b) block diagram with equivalent transformations.

However, Figure 3.5 has $\tilde{X}^{-1}(s)$ in several blocks, and since $Q(s)$ cannot be treated alone, we tried to transform it and obtained Figure 3.8a [4, 5]. Furthermore, using the double coprime factorization $\tilde{X}(s)D(s) + \tilde{Y}(s)N(s) = I$ of Equation (3.19), we can transform it into Figure 3.8b. The block diagram is thus simplified, and the representation with $Q(s)$ outside is often used. Based on this structure, the disturbance observer could be designed according to robust control theory, for example, by assuming that $Q(s)$ is not for disturbance rejection, but varying the control plant $P(s)$, so that the stability condition is satisfied. However, this is beyond the scope of this book. Interested readers are referred to the literature such as [4–7].

References

1 Kemin Zhou, John C. Doyle, Keith Glover: Robust and Optimal Control, Prentice Hall, 1995.
2 Bruce A. Francis: A Cource in H_∞ Control Theory, Lecture Notes in Control and Information Sciences, Springer-Verlag, 1987.

3 Shiro Masuda, Valeri T. Kroumov, Akira Inoue, Kenji Sugimoto: A Design Method of Multivariable Model Reference Adaptive Control System Using Coprime Factrization Approach, Proceedings of the 32nd Automatic Control Conference on Decision and control, FM4-3:10, 1993.

4 Kazunori Matsumoto Tatsuya Suzuki Shigeru Okuma: The Design Method of Two-Degree-of-Freedom Controller Using μ-Synthesis and Its Application to Two-Mass System, IEEE 4th International Workshop on Advanced Motion Control (AMC'96), 606–611, 1996.

5 Akitaka Mizutani, Kazuhiro Yubai, Junji Hirai: A Direct Design from Input/Output Data of the Youla Parameter for Compensating Plant Perturbation on GIMC Structure, IEEE 35th Annual Conference of IEEE Industrial Electronics, 3047–3052, 2009.

6 Kiyoshi Ohishi, Toshimasa Miyazaki, Yoshihiro Nakamura: High performance ultra-low speed servo system based on doubly coprime factorization and instantaneous speed observer, IEEE Transactions on Mechatronics, Vol. 1, No. 1, 89–98, 1996.

7 Tstomu Mita, Mitsuo Hirata, Kenichi Murata, Hui Zhang: H_∞ control versus disturbance-observer-based control, IEEE Transactions on Industrial Electronics, Vol. 45, No. 3, 488–495, 1998.

4

Disturbance Observer in State Space

This chapter introduces how to design disturbance observers (DOBs) in continuous-time systems, following the basic observer theory of modern control theory.

4.1 Identity Input Disturbance Observer

4.1.1 How to Design the Identity Input Disturbance Observer in Continuous System

The **state** and **output** equations with input disturbances are represented in Equations (4.1) and (4.2):

$$\dot{x}(t) = Ax(t) + Bu(t) - Bd(t) \tag{4.1}$$

$$y(t) = Cx(t) \tag{4.2}$$

where state variables $x \in R^n$, control input and input disturbance $u, d \in R^m$, observed output $y \in R^l$, and coefficient matrices $A \in R^{n \times n}, B \in R^{n \times m}$, and $C \in R^{l \times n}$. Assuming that the disturbance does not change over a short period, d satisfies $\dot{d} = 0_{m \times 1}$. The d is also added to the state variables, $\bar{x} = [x, d]^T$, creating the **extended system**.

$$\begin{bmatrix} \dot{x} \\ \dot{d} \end{bmatrix} = \begin{bmatrix} A & -B \\ 0 & 0 \end{bmatrix} \begin{bmatrix} x \\ d \end{bmatrix} + \begin{bmatrix} B \\ 0 \end{bmatrix} u \tag{4.3}$$

$$y = \begin{bmatrix} C & 0 \end{bmatrix} \begin{bmatrix} x \\ d \end{bmatrix} \tag{4.4}$$

Let the coefficient matrices of the first and second terms on the right-hand side of Equation (4.3) be $\bar{A}$ and B, respectively, and the coefficient matrix of the right-hand side of Equation (4.4) be $\bar{C}$. The above equation can be expressed as $\dot{\bar{x}} = \bar{A}\bar{x} + \bar{B}u$

Disturbance Observer for Advanced Motion Control with MATLAB/Simulink, First Edition. Akira Shimada.
© 2023 The Institute of Electrical and Electronics Engineers, Inc. Published 2023 by John Wiley & Sons, Inc.
Companion website: www.wiley.com/go/disturbanceobserver

and $y = \bar{C}\bar{x}$. If the extended system $[\bar{A}, \bar{C}]$ is observable, then the observer gain matrix H can be designed.[1] The **identity disturbance observer** can be obtained by dividing H into $H = [H_1^T, H_2^T]^T$.

$$\begin{bmatrix} \dot{\hat{x}} \\ \dot{\hat{d}} \end{bmatrix} = \begin{bmatrix} A & -B \\ 0 & 0 \end{bmatrix} \begin{bmatrix} \hat{x} \\ \hat{d} \end{bmatrix} + \begin{bmatrix} B \\ 0 \end{bmatrix} u + \begin{bmatrix} H_1 \\ H_2 \end{bmatrix} (y - \hat{y}) \tag{4.5}$$

$$\hat{y} = \begin{bmatrix} C & 0 \end{bmatrix} \begin{bmatrix} \hat{x} \\ \hat{d} \end{bmatrix} \tag{4.6}$$

Figure 4.1 represents it graphically. This feature is that it can estimate the original state variable x and the disturbance d simultaneously.

Examining Figure 4.1 shows that the dimensions of the upper control plant model and the lower observer are different. The original state variable estimate $\hat{x}$ and the disturbance estimate $\hat{d}$ are included in the expanded state variable estimate $\hat{\bar{x}}$. It seems inconvenient to extract and use one of the estimates from it. Therefore, based on the structure of Equations (4.5) and (4.6), we can reexpress $\hat{x}$ and $\hat{d}$ separately to obtain Figure 4.2. In this figure, the control plant and the observer are represented by the same structure and dimension, which is easy to understand.

For reference, the block diagram for the case where $\dot{x}(t) = Ax(t) + Bu(t) - B_{dis}(t)$, but $B \neq B_{dis}$, is shown in Figure 4.3. In this case, the $\bar{A}$ matrix of the extended

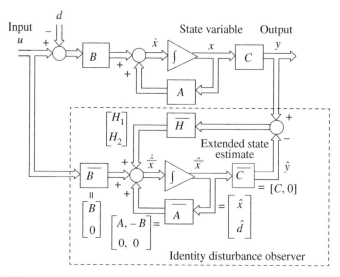

Figure 4.1 Block diagram of the identity disturbance observer.

---

1 See the appendix for more on observability.

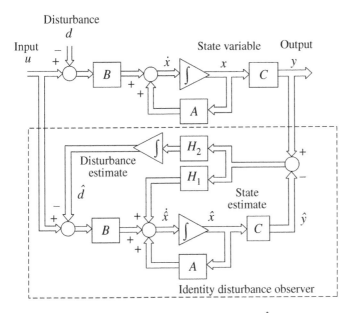

Figure 4.2 Identity disturbance observer with $\hat{x}$ and $\hat{d}$ represented separately.

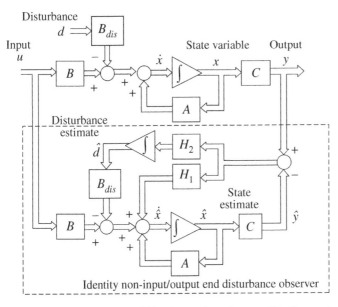

Figure 4.3 Block diagram of identity noninput/output disturbance observer.

system is $\begin{bmatrix} A & -B_{dis} \\ 0 & 0 \end{bmatrix}$. The $\overline{C}$ is not replaced. If $(\overline{A}, \overline{C})$ is observable, we can still design a DOB. However, in the following continuous DOBs, we mainly use the form in Figure 4.2. The following example shows how to use it.

Example 4.1.1 (Identity input disturbance observer with velocity observation.) Consider a specific design method assuming that the control plant is a mechatronics system with a velocity sensor. That is, let us try to design the cart's identity DOB which can observe the velocity. In Equation (2.1), $x(t) = v_c(t)$, $u(t) = f(t)$, and $y(t) = v_c(t)$, and we have $\dot{x}(t) = \dot{v}_c = \frac{1}{m}f(t) - \frac{1}{m}d(t)$, $y(t) = v_c(t)$ Thus, we get $A = 0$, $B = \frac{1}{m}$, and $C = 1$. Then, setting $\overline{x}(t) = [v_c(t), d(t)]^T$, we can express the extended system as follows:

$$\dot{\overline{x}}(t) = \begin{bmatrix} \dot{v}_c(t) \\ \dot{d}(t) \end{bmatrix} = \begin{bmatrix} 0 & -\frac{1}{m} \\ 0 & 0 \end{bmatrix} \begin{bmatrix} v_c(t) \\ d(t) \end{bmatrix} + \begin{bmatrix} \frac{1}{m} \\ 0 \end{bmatrix} f(t)$$

$$y(t) = v_c(t) = \begin{bmatrix} 1 & 0 \end{bmatrix} \begin{bmatrix} v_c(t) \\ d(t) \end{bmatrix}$$

From this, the DOB is Equations (4.7) and (4.8):

$$\begin{bmatrix} \dot{\hat{v}}_c \\ \dot{\hat{d}} \end{bmatrix} = \begin{bmatrix} 0 & -\frac{1}{m} \\ 0 & 0 \end{bmatrix} \begin{bmatrix} \hat{v}_c \\ \hat{d} \end{bmatrix} + \begin{bmatrix} \frac{1}{m} \\ 0 \end{bmatrix} f(t) + \begin{bmatrix} H_1 \\ H_2 \end{bmatrix} (v_c - \hat{v}_c)$$

$$\tag{4.7}$$

$$\hat{y} = \begin{bmatrix} 1 & 0 \end{bmatrix} \begin{bmatrix} \hat{v}_c \\ \hat{d} \end{bmatrix}$$

$$\tag{4.8}$$

If we set the poles of the observer as p_1, p_2, then

$$\det(\lambda I - (\overline{A} - H\overline{C})) = \left| \lambda I_2 - \left(\begin{bmatrix} 0 & -\frac{1}{m} \\ 0 & 0 \end{bmatrix} - \begin{bmatrix} H_1 \\ H_2 \end{bmatrix} \begin{bmatrix} 1 & 0 \end{bmatrix} \right) \right|$$

$$= \begin{vmatrix} \lambda + H_1 & \frac{1}{m} \\ H_2 & \lambda \end{vmatrix} = \lambda^2 + H_1 \lambda - H_2/m$$

$$= (\lambda - p_1)(\lambda - p_2) = \lambda^2 - (p_1 + p_2)\lambda + p_1 p_2 \tag{4.9}$$

By comparing the coefficients, we obtain $H_1 = -(p_1 + p_2)$ and $H_2 = -m p_1 p_2$. The corresponding example program is shown in List 4.1.[2]

---

2 The observer gain calculation uses the place function instead of the coefficient comparison. The transpose of $A_{bar}, C_{bar}, A'_{bar}, C'_{bar}$, was used. Though the state feedback $u = -Fx$ to $\dot{x} = Ax + Bu.F$, which corresponds to the eigenvalues of $\dot{x} = Ax + B(-Fx) = (A - BF)x$, whereas H corresponding to the eigenvalues of the time derivative of the estimated deviation $e = x - \hat{x}$, $\dot{e} = (A - HC)e$, was calculated with the duality, in contrast that $[A, B]$ is controllable, $[A, C]$ is observable. It is equivalent that $[AT, CT]$ is controllable.

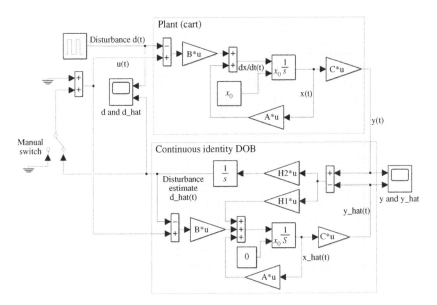

Figure 4.4 Simulink model of velocity observation input disturbance observer.

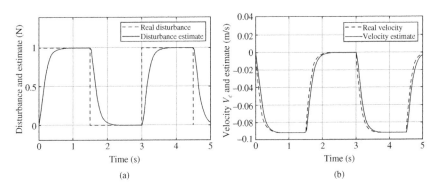

Figure 4.5 Simulated waveform of a velocity observation input disturbance observer. (a) Disturbance and disturbance estimates and (b) velocity and velocity estimates.

Figure 4.4 is an example of Simulink, and Figure 4.5a,b are the result of estimation of disturbance and velocity.

List 4.1:   Identity input disturbance observer (with velocity observable).

1	%% Physical Parameters
2	m=2; % Mass[kg]
3	dis_step=1.0;
4	%% State space model

```
5   A=0;B=1/m;C=1;% ABC of continuous system
6
7   %% Creating an expanded system and designing the observer gain
8   Abar=[A,-B;0,0];Bbar=[B;0];Cbar=[C,0];
9   Ov=[Cbar;Cbar*Abar;Cbar*Abar^2];rank_of_Ov=rank(Ov);% Evaluate rank
10  pole_ob=[-10,-12]; % POle of DOB
11  Htemp=place(Abar',Cbar',pole_ob);H=Htemp';H1=H(1);H2=H(2);
12
13  %% Simulation
14  x0=[0.0];tend=5.0; % Initial velocity and simulation time
15  open_system('sim_Figure_4_4_and_4_5_dob_input_dis_vel_out');
16  set_param('sim_Figure_4_4_and_4_5_dob_input_dis_vel_out','WideLines','on');
17  set_param('sim_Figure_4_4_and_4_5_dob_input_dis_vel_out','ShowLineDimensions','on');
18  z=sim('sim_Figure_4_4_and_4_5_dob_input_dis_vel_out');
```

In this way, it is possible to estimate the velocity considering the disturbance and the external disturbance. In this example, it took about 0.5 seconds to estimate the square wave-shaped disturbance, but the estimated velocity can be changed by adjusting the poles of the observer.

Example 4.1.2 (Identity input disturbance observer with position observation.) Consider the case where the observation output is position information. If $x(t) = [x_c(t), v_c(t)]^T$, $u(t) = f(t)$, and $y(t) = x_c(t)$, the state and output equations are Equations (4.10) and (4.11):

$$\begin{bmatrix} \dot{x}_c \\ \dot{v}_c \end{bmatrix} = \begin{bmatrix} 0 & 1 \\ 0 & 0 \end{bmatrix} \begin{bmatrix} x_c \\ v_c \end{bmatrix} + \begin{bmatrix} 0 \\ \frac{1}{m} \end{bmatrix} f - \begin{bmatrix} 0 \\ \frac{1}{m} \end{bmatrix} d \qquad (4.10)$$

$$y = \begin{bmatrix} 1 & 0 \end{bmatrix} \begin{bmatrix} x_c \\ v_c \end{bmatrix} \qquad (4.11)$$

Next, we create an extended system by rearranging $\bar{x} = [x^T, d^T]^T = [x_c, v_c, d]^T$ and design a DOB as in the previous example, which is expressed as Equations (4.12) and (4.13):

$$\begin{bmatrix} \dot{\hat{x}}_c \\ \dot{\hat{v}}_c \\ \dot{\hat{d}} \end{bmatrix} = \begin{bmatrix} 0 & 1 & 0 \\ 0 & 0 & -\frac{1}{m} \\ 0 & 0 & 0 \end{bmatrix} \begin{bmatrix} \hat{x}_c \\ \hat{v}_c \\ \hat{d} \end{bmatrix} + \begin{bmatrix} 0 \\ \frac{1}{m} \\ 0 \end{bmatrix} f + \begin{bmatrix} H_{11} \\ H_{12} \\ H_2 \end{bmatrix} (x_c - \hat{x}_c) \qquad (4.12)$$

$$\hat{y} = \begin{bmatrix} 1 & 0 & 0 \end{bmatrix} \begin{bmatrix} \hat{x}_c & \hat{v}_c & \hat{d} \end{bmatrix}^T \qquad (4.13)$$

For the observer gain $H = [H_1^T, H_2^T]^T = [H_{11}, H_{12}, H_2]^T$, choose the observer poles p_1, p_2, and p_3 as above, find the eigenvalues of $\bar{A} - H\bar{C}$, and compare the coefficients.

$$det(\lambda I - (\bar{A} - H\bar{C}))$$

$$= \left| \lambda I_3 - \left(\begin{bmatrix} 0 & 1 & 0 \\ 0 & 0 & -\frac{1}{m} \\ 0 & 0 & 0 \end{bmatrix} - \begin{bmatrix} H_{11} \\ H_{12} \\ H_2 \end{bmatrix} [1 \ 0 \ 0] \right) \right|$$

$$= \left| \begin{matrix} \lambda + H_{11} & -1 & 0 \\ H_{12} & \lambda & \frac{1}{m} \\ H_2 & 0 & \lambda \end{matrix} \right| = \lambda^3 + H_{11}\lambda^2 + H_{12}\lambda - H_2/m$$

$$= (\lambda - p_1)(\lambda - p_2)(\lambda - p_3)$$

$$= \lambda^3 - (p_1 + p_2 + p_3)\lambda^2 + (p_1 p_2 + p_2 p_3 + p_3 p_1)\lambda - p_1 p_2 p_3 \qquad (4.14)$$

By comparing the coefficients, we obtain $H_{11} = -(p_1 + p_2 + p_3)$, $H_{12} = p_1 p_2 + p_2 p_3 + p_3 p_1$, and $H_2 = m p_1 p_2 p_3$.

List 4.2 shows an example program for comparing the return of the disturbance estimates with and without the return of the disturbance estimates with the identity input DOB for positional observations and a comparison with a general identity observer. An example of the Simulink model configuration is shown in Figure 4.6.

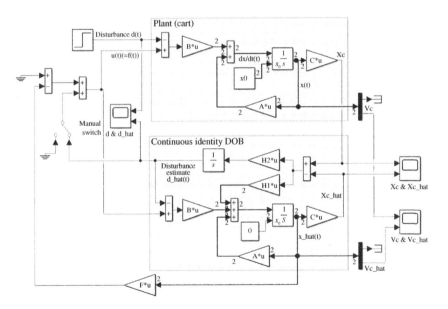

Figure 4.6 Simulink model of control system with input disturbance observer and position observation.

List 4.2:   Identity input disturbance observer (with positional observation).

```
1    %% Physical Parameters
2    m=2; % Mass[kg]
3    dis_step=10.0; % Magnitude of disturbance [N]
4    %% State space model
5    A=[0,1;0,0];B=[0;1/m];C=[1 0];% A,B,C matrices
6
7    %% Creating an expanded system and designing the observer gain
8    Abar=[A,-B;zeros(1,2),0];Bbar=[B;0];Cbar=[C,0]; % Create extended system
9    Ov=[Cbar;Cbar*Abar;Cbar*Abar^2];rank_of_Ov=rank(Ov);% Confirmation of rank
10   pole_dob=[-13,-14,-15]*5; % Poles of DOB
11   Htemp=place(Abar',Cbar',pole_dob);H=Htemp';H1=H(1:2);H2=H(3);
12
13   %% State feedback gain design with LQ control
14   Q=diag([1e3,1e1]);R=1e2;F=lqr(A,B,Q,R);
15
16   %% Simulation
17   x0=[0.1;0.0];tend=10.0; % Initial velocity and simulation time
18   open('sim_Figure_4_6_and_4_7_dob_input_dis_pos_out');
19   set_param('sim_Figure_4_6_and_4_7_dob_input_dis_pos_out','WideLines','on');
20   set_param('sim_Figure_4_6_and_4_7_dob_input_dis_pos_out','ShowLineDimensions','on');
21   y=sim('sim_Figure_4_6_and_4_7_dob_input_dis_pos_out');
22
23   %% Normal observer design used to compare with DOB
24   % The comparison result is shown in Figure 4.8.
25   pole_ob=[-13,-15]*5;
26   H_temp_=place(A',C',pole_ob);Hnormal=H_temp_';
27   %
28   % Simulation with normal observer used to compare with DOB
29   % The comparison result is shown in Figure 4.8.
30   % open_system('sim_Figure_4_8_ob_input_dis_pos_out');
31   z=sim('sim_Figure_4_8_ob_input_dis_pos_out');
```

Figure 4.7a-1 shows the waveforms of the disturbance and disturbance estimate. Figure 4.7a-2 shows the results of velocity and velocity estimation when the Manual Switch is set to ON (right side) and the positive feedback of the disturbance estimate. Figure 4.7b-2 shows the results of velocity and velocity estimation when the Manual Switch is set to ON(right side). The figure shows the results of velocity and velocity estimation when Manual Switch is set to OFF(left side), and no positive feedback of the disturbance estimate is performed. With OFF, there is a large fluctuation in velocity due to the effect of the disturbance, but the estimated velocity is consistent with the velocity waveform, indicating that the estimation works well.

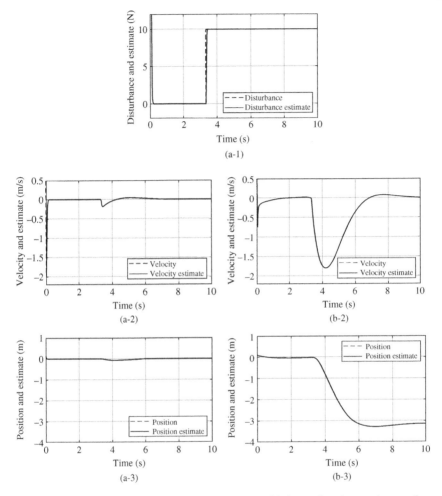

Figure 4.7 Simulation results of the control system with input disturbance observer for position observation. (a-1) Disturbances and disturbance estimates, (a-2) velocity and estimate (SW on), (a-3) position and estimate (SW on), (b-2) velocity and estimate (SW off), and (b-3) position and estimate (SW off).

Figure 4.7a-3 shows the position estimation results at ON, and (b-3) shows the position and estimation results for OFF. Although the position and velocity change due to the disturbance, they remain at a small value when ON. What would happen to the disturbance estimate if the disturbance is time varying instead of step? Some people worry that the disturbance cannot be estimated because it is designed assuming $\dot{d} = 0$. However, time-varying disturbances can be estimated because the observer has dynamics and feeds back the output estimation error through

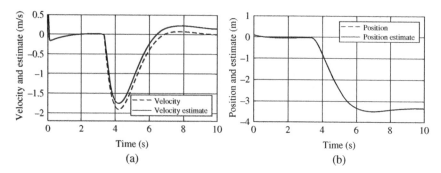

Figure 4.8 Simulation of the control system with identity observer in position observation (for comparison). (a) Velocity and estimated value and (b) position and estimated value.

the observer gain $H = [H_1^T, H_2^T]^T$. If a large observer gain cannot be used due to noise or other reasons, the design assuming higher order disturbances becomes effective.

For comparison, a simulation example using a general identity observer is presented in Figure 4.8.

The velocity estimation waveform in Figure 4.8a follows the velocity waveform but is off. The result is similar to the waveform in Figure 4.7b-2, and Figure 4.8b is almost identical to Figure 4.7b-3. These results show that the identity DOB and the disturbance estimation function accurately estimate the physical quantities other than the disturbance, regardless of whether the Manual Switch is set to ON or OFF. Is there a case where SW OFF is actively used in a control system with an input DOB? Various unexpected things can happen in real systems. For example, the positive feedback of the disturbance estimate by ON sometimes stimulates an unobservable part of the existing mechatronics system causing the noise. Or the modeling errors sometimes affect the system causing vibrations. A typical example is introduced in Section 6.2.

We can choose SW OFF and leave the disturbance suppression to an external compensator to avoid degrading the velocity estimation performance with a slow estimating observer.

4.1.2 Controllability and State Feedback

State feedback is the basis of control in modern control theory. For this to be possible, the control plant must be controllable. Generally, DOBs are used for controllable and observable control plants, but is an extended system controllable? If obeying the definition of controllability, each element of the extended state variable $\bar{x}(t) = [x^T(t); d^T(t)]^T$ can converge to an arbitrary final value $\bar{x}e$ in a finite time by using an adequate control input. In a single-input system, we find that $det(U_c) = det[\bar{B}, \bar{A}\bar{b}, \ldots, \bar{A}^{n-1}\bar{B}] = 0$. This means that $rank(U_c) = 0$, an irreducible control.[3] Therefore, the controllable control plant $P(s) = \{A, B, C, D\}$ is stabilized by feedback control, and, if necessary, the disturbance estimate $\hat{d}$ is returned. If it is controllable, an adequate control input converges the extended state variable $\bar{x}(t) = [x^T(t); d^T(t)]^T$ to an arbitrary final value $\bar{x}e$ in a finite time. However, we cannot changes unknown disturbance artificially. So the extended system is uncontrollable even without checking the controllability. Therefore, we stabilize the controllable control plant with some feedback control usually, and if necessary, make use of the disturbance estimate $\hat{d}$.

4.1.3 Continuous-time Servo System with Identity Disturbance Observer

The design procedure for the servo system with a DOB is also shown. Please refer to Appendix A.3 for the basic design method.

Example 4.1.3 (Servo system with identity disturbance observer with position observation.)   The physical quantities, state equations, and output equations are the same as in Equations (4.15) and (4.16). Let x_{ref} be the position reference value, and $e(t) = \dot{w}(t) = x_{ref} - x_c(t)$ the position deviation. Assume that $\dot{x}_{ref} = 0$ and $\dot{d} = 0$. Putting the extended system as $\tilde{x}(t) = [x^T(t), w^T(t)]^T = [x_c(t), v_c(t), w(t)]^T$, we can define it as in Equations (4.15) and (4.16) as follows:

$$\begin{bmatrix} \dot{x}_c(t) \\ \dot{v}_c(t) \\ \dot{w}(t) \end{bmatrix} = \begin{bmatrix} 0 & 1 & 0 \\ 0 & 0 & 0 \\ -1 & 0 & 0 \end{bmatrix} \begin{bmatrix} x_c(t) \\ v_c(t) \\ w(t) \end{bmatrix} + \begin{bmatrix} 0 \\ \frac{1}{m} \\ 0 \end{bmatrix} \{f(t) - d\}$$
$$+ \begin{bmatrix} 0 & 0 & 1 \end{bmatrix}^T x_{ref} \tag{4.15}$$

$$y = \begin{bmatrix} 1 & 0 & 0 \end{bmatrix} \begin{bmatrix} x_c(t) & v_c(t) & w(t) \end{bmatrix}^T \tag{4.16}$$

Define Equation (4.17) in $t \to \infty$:

$$\begin{bmatrix} \dot{x}_c(\infty) \\ \dot{v}_c(\infty) \\ \dot{w}(\infty) \end{bmatrix} = \begin{bmatrix} 0 & 1 & 0 \\ 0 & 0 & 0 \\ -1 & 0 & 0 \end{bmatrix} \begin{bmatrix} x_c(\infty) \\ v_c(\infty) \\ w(\infty) \end{bmatrix} + \begin{bmatrix} 0 \\ \frac{1}{m} \\ 0 \end{bmatrix} \{f(\infty) - d\}$$
$$+ \begin{bmatrix} 0 & 0 & 1 \end{bmatrix}^T x_{ref} \tag{4.17}$$

---

3 0s line up in the bottom line of the U_c. Consequently, the $det(Uc) = 0$, which does not satisfy the rank condition for controllable control as shown in the appendix, and therefore, it is uncontrollable.

$x_e = x_c(t) - x_c(\infty)$, $v_e = v_c(t) - v_c(\infty)$, $w_e = w(t) - w(\infty)$, and $f_e = f(t) - f(\infty)$, and taking the difference between the two sides of Equations (4.15) and (4.17), we get Equation (4.18):

$$\begin{bmatrix} \dot{x}_e \\ \dot{v}_e \\ \dot{w}_e \end{bmatrix} = \begin{bmatrix} 0 & 1 & 0 \\ 0 & 0 & 0 \\ -1 & 0 & 0 \end{bmatrix} \begin{bmatrix} x_e \\ v_e \\ w_e \end{bmatrix} + \begin{bmatrix} 0 \\ \frac{1}{m} \\ 0 \end{bmatrix} f_e$$

(4.18)

Let the feedback gain $\tilde{F} = [F_1, F_2]$ be obtained and $F_1 = [F_{11}, F_{12}]$. If we write $f_e = f(t) - f(\infty) = -[F_{11}, F_{12}, F_2][x_c(t) - x_c(\infty), v_c(t) - v_c(\infty), w(t) - w(\infty)]^T$ element by element, then $f(t) = -F_{11}x_c(t) - F_{12}v_c(t) - F_2w(t) + \{f(\infty) + F_{11}x_c(\infty) + F_{12}v_c(\infty) + F_2w(\infty)\}$, $F = F_1$, and $K_i = -F_2$ and replace $F(\infty) + F_{11}x_c(\infty) + F_{12}v_c(\infty) + F_2w(\infty) = 0$, replace $x(t)$ with $\hat{x}(t)$, and organize the control input including the disturbance estimate $\hat{d}(t)$ to obtain Equation (4.19):

$$f(t) = -F\hat{x}(t) + K_iw(t) + \hat{d}(t)$$

$$= -[F_{11}, F_{12}] \begin{bmatrix} \hat{x}_c(t) \\ \hat{v}_c(t) \end{bmatrix} + K_iw(t) + \hat{d}(t)$$

(4.19)

Put the servo system program example in List 4.3 and the Simulink model in Figure 4.9. The simulation results are shown in Figure 4.10. The servo gain $\tilde{F}$ is calculated using the MATLAB 'lqr' command as the LQ servo system.[4]

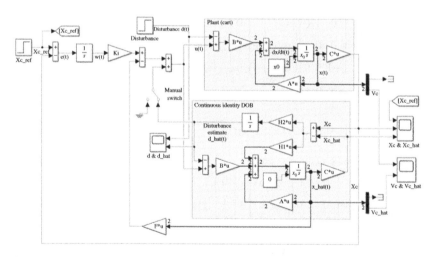

Figure 4.9 Simulink model of identity DOB-combined control system with position observation.

---

4 See Appendix about the detail of LQ control.

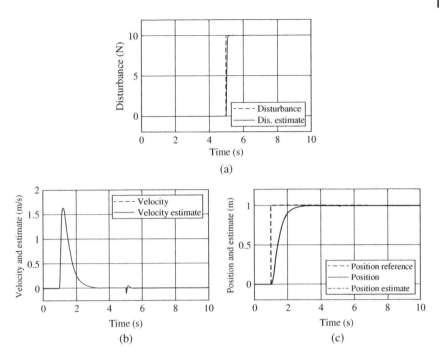

Figure 4.10 Simulation of a servo system with input end DOB with position observation. (a) Disturbance and disturbance estimate, (b) velocity and estimated velocity value, and (c) position and estimated position value.

The disturbance estimation function is described in the example above. The disturbance estimation function is the same as in the previous example.

List 4.3: Servo system with identity input disturbance observer (with position observation).

```
1   %% Physical Parameters
2   m=2; % Mass[kg]
3   dis_step=10.0; % Magnitude of disturbance [N]
4   %% State space model
5   A=[0,1;0,0];B=[0;1/m];C=[1 0];% A,B,C matrices
6
7   %% Continuous identity DOB design
8   Abar=[A,-B;zeros(1,2),0];Bbar=[B;0];Cbar=[C,0]; % Create extended system
9   Ov=[Cbar;Cbar*Abar;Cbar*Abar^2];rank_of_Ov=rank(Ov); % Confirmation of rank
10  pole_dob=[-13,-14,-15]*5; % Pole of DOB
11  Htemp=place(Abar',Cbar',pole_dob);H=Htemp';H1=H(1:2);H2=H(3);
12
13  %% Servo system design
```

```
14   Atilde=[A,zeros(2,1);−C,0];Btilde=[B;0];
15   Q=diag([3e5,3e3,2e6]);R=1e0;Ftilde=lqr(Atilde,Btilde,Q,R);
16   F=Ftilde(1:2);Ki=−Ftilde(3);
17
18   %% Simulation
19   x0=[0.0;0.0];tend=10.0;Xc_ref=1.0; % Initial velocity and simulation time, reference
20   open_system('sim_Figure_4_9_and_4_10_servo_with_dob_pos_out');
21   set_param('sim_Figure_4_9_and_4_10_servo_with_dob_pos_out','WideLines','on');
22   set_param('sim_Figure_4_9_and_4_10_servo_with_dob_pos_out','ShowLineDimensions','on');
23   z=sim('sim_Figure_4_9_and_4_10_servo_with_dob_pos_out'); % Run servo system with dob
```

The effect of the disturbance is slightly visible in Figure 4.10b,c immediately after five seconds, but the position converges nicely to the reference value.

4.2 Identity Reaction Force Observer

If there are elements that are represented by linear systems, such as viscous friction or elasticity, should they be included in the model to be controlled, or should they be included in the disturbance? It is up to the designer to decide whether to include them in the disturbance or in the model to be controlled. It seems more natural to include nonlinear elements such as Coulomb friction in the disturbance. However, when estimating the reaction force, if it is included in the disturbance, it will be estimated as a disturbance including other elements. It is difficult to extract only the reaction force. In this case, a method of transforming the input can be considered.

Example 4.2.1 (Reaction force estimation observer with position observation.) We attempt to design an identity reaction force observer for a cart with position observation. In Equation (2.5), let $\bar{f}(t) = f(t) - cv_c(t) - f_c\,\text{sign}(v_c(t))$ be the apparent input, and considering Equation (2.6), Equations (4.20) and (4.21) can be obtained, and the reaction force observer equations (4.22) and (4.23) can be designed through the extended system.

$$\begin{bmatrix} \dot{x}_c \\ \dot{v}_c \end{bmatrix} = \begin{bmatrix} 0 & 1 \\ 0 & 0 \end{bmatrix}\begin{bmatrix} x_c \\ v_c \end{bmatrix} + \begin{bmatrix} 0 \\ \frac{1}{m} \end{bmatrix}\bar{f} - \begin{bmatrix} 0 \\ \frac{1}{m} \end{bmatrix}d$$

$$= \begin{bmatrix} 0 & 1 \\ 0 & 0 \end{bmatrix}\begin{bmatrix} x_c \\ v_c \end{bmatrix} + \begin{bmatrix} 0 \\ \frac{1}{m} \end{bmatrix}\{f - cv_c - f_c\,\text{sign}(v_c)\}$$

$$- \begin{bmatrix} 0 \\ \frac{1}{m} \end{bmatrix}d \tag{4.20}$$

$$y = \begin{bmatrix} 1 & 0 \end{bmatrix}\begin{bmatrix} x_c \\ v_c \end{bmatrix} \tag{4.21}$$

$$
\begin{bmatrix} \dot{\hat{x}}_c \\ \dot{\hat{v}}_c \\ \dot{\hat{d}} \end{bmatrix} = \begin{bmatrix} 0 & 1 & 0 \\ 0 & 0 & -\frac{1}{m} \\ 0 & 0 & 0 \end{bmatrix} \begin{bmatrix} \hat{x}_c \\ \hat{v}_c \\ \hat{d} \end{bmatrix} + \begin{bmatrix} 0 \\ \frac{1}{m} \\ 0 \end{bmatrix} \{ f - c\hat{v}_c - f_c \operatorname{sign}(\hat{v}_c) \}
$$

$$
+ \begin{bmatrix} H_{11} \\ H_{12} \\ H_2 \end{bmatrix} (y - \hat{y})
$$

(4.22)

$$
\hat{y} = \begin{bmatrix} 1 & 0 & 0 \end{bmatrix} \begin{bmatrix} \hat{x}_c & \hat{v}_c & \hat{d} \end{bmatrix}^T
$$

(4.23)

where we use the force $\bar{f}(t) = f(t) - cv_c(t) - f_c \operatorname{sign}(v_c(t))$ instead of $\bar{f}(t) = f(t) - c\hat{v}_c(t) - f_c \operatorname{sign}(\hat{v}_c(t))$ within the observer. The viscous friction force and Coulomb friction differ from the situation in Section 2.4 because they use velocity estimates. An example program is shown in List 4.4, and the Simulink model and estimated waveforms are shown in Figures 4.11 and 4.12.

List 4.4: Example of identity reaction force estimation observer (position observation type).

```
1   %% Physical Parameters
2   m=2;c=0.5; % Mass[kg]& coefficient of viscous friction[N/(m/s)]
3   k=15.0;fc=3.0;% Modulus of elasticity[N/m]& magnitude of Coulomb friction[N]
4   %% State space model
5   A=[0,1;0,0];B=[0;1/m];C=[1 0]; % A,B,C matrices created ignoring k and c
6
7   %% Create extended system and observer gain design
8   Abar=[A,-B;zeros(1,2),0];Bbar=[B;0];Cbar=[C,0];
9   Ov=[Cbar;Cbar*Abar;Cbar*Abar^2];rank_of_Ov=rank(Ov);% Confirmation of rank
10  % DOB design
11  pole_ob=[-20,-22,-24]; % Poles of DOB
12  Htemp=place(Abar',Cbar',pole_ob);H=Htemp';H1=H(1:2);H2=H(3);
13  % RFOB design
14  pole_rfob=[-30,-32,-34]; % Poles of RFOB
15  Hrtemp=place(Abar',Cbar',pole_rfob);Hr=Hrtemp';Hr1=Hr(1:2);Hr2=Hr(3);
16
17  %% State feedback controller design
18  pole=[-3,-4];F=place(A,B,pole);
19
20  %% Simulation
21  x0=[0.1;0.0];tend=5.0; % Initial values of position & velocity and simulation time
22  open_system('sim_Figure_4_11_and_4_12_continuous_RFOB_pos');
23  set_param('sim_Figure_4_11_and_4_12_continuous_RFOB_pos','WideLines','on');
24  set_param('sim_Figure_4_11_and_4_12_continuous_RFOB_pos','ShowLineDimensions','on');
25  z=sim('sim_Figure_4_11_and_4_12_continuous_RFOB_pos');
```

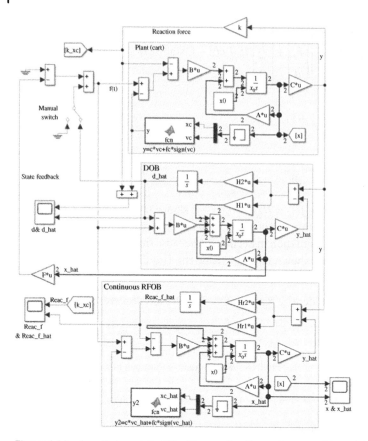

Figure 4.11 Simulink model of identity reaction force observer with positional observation.

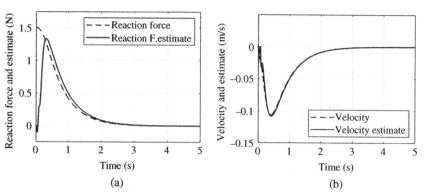

Figure 4.12 Simulation results of identity reaction force observer with position observation. (a) Reaction force and reaction force estimate and (b) velocity and velocity estimate.

In this example, the upper part of the Simulink model is the cart model to be controlled, and the reaction force k_xc is returned with a negative sign. The middle part is the identity DOB, and the disturbance estimate d_hat is returned with a positive sign via Manual Switch. Additionally, the state feedback $f(t) = -F\hat{x}(t)$ stabilizes the system using the feedback gain F obtained through the pole assignment method, and the position is converged to the origin. Assume that the initial value $x(0)$ is known by both observers.

4.3 Identity Output Disturbance Observer

It is possible to design an observer that estimates disturbances to the observed output. Nevertheless, what is output disturbance? In the case of a mechatronic system, the main disturbance may be the offset or fluctuation of the sensor signal, or it may include a shift in the mounting position of the sensor due to the extension or contraction of the mechanism.

Let us assume that the observed output $y = Cx - d_o$, but $\dot{d}_o = 0$. The equation of state is the general equation (4.24), and the output equation, including the disturbance to the output end, is Equation (4.25).

$$\dot{x}(t) = Ax(t) + Bu(t) \qquad (4.24)$$

$$y(t) = Cx(t) - d_o(t) \qquad (4.25)$$

The extended system is represented as follows:

$$\begin{bmatrix} \dot{x} \\ \hline \dot{d}_o \end{bmatrix} = \begin{bmatrix} A & 0 \\ \hline 0 & 0 \end{bmatrix} \begin{bmatrix} x \\ \hline d \end{bmatrix} + \begin{bmatrix} B \\ \hline 0 \end{bmatrix} u \qquad (4.26)$$

$$y = \begin{bmatrix} C & -I \end{bmatrix} \begin{bmatrix} x \\ \hline d_o \end{bmatrix} \qquad (4.27)$$

If Equations (4.24) and (4.25) are observable, we can design the observer gain matrix H_o. Dividing the contents of H_o as $H_o = [H_{o1}^T, H_{o2}^T]^T$, we obtain the identity output DOBs, Equations (4.28) and (4.30).

$$\begin{bmatrix} x \\ \hline d_o \end{bmatrix} = \begin{bmatrix} A & 0 \\ \hline 0 & 0 \end{bmatrix} \begin{bmatrix} x \\ \hline d_o \end{bmatrix} + \begin{bmatrix} B \\ \hline 0 \end{bmatrix} u$$

$$+ \begin{bmatrix} H_{o1} \\ \hline H_{o2} \end{bmatrix} (y - \hat{y}) \qquad (4.28)$$

$$\hat{y} = [C \vdots -I] \begin{bmatrix} \hat{x} \\ \hline \hat{d}_o \end{bmatrix}$$

(4.29)

$$\hat{d}_o = y - C\hat{x}$$

(4.30)

Example 4.3.1 (Identity output disturbance observer with position observation.) Attempt to simulate the cart used in the input disturbance assuming an output step disturbance d_o. Consequently, it was unobservable, as shown below. First, the equation of motion is expressed as follows:

$$m\dot{v}_c + kx_c = f$$

$$y = x_c - d_o$$

The corresponding state and output equations are Equations (4.31) and (4.32):

$$\begin{bmatrix} \dot{x}_c \\ \dot{v}_c \end{bmatrix} = \begin{bmatrix} 0 & 1 \\ -k/m & 0 \end{bmatrix} \begin{bmatrix} x_c \\ v_c \end{bmatrix} + \begin{bmatrix} 0 \\ \frac{1}{m} \end{bmatrix} f$$

(4.31)

$$y = \begin{bmatrix} 1 & 0 \end{bmatrix} \begin{bmatrix} x_c \\ v_c \end{bmatrix} - d_o$$

(4.32)

The extended system is given by Equations (4.33) and (4.34):

$$\begin{bmatrix} \dot{x}_c \\ \dot{v}_c \\ \hline \dot{d}_o \end{bmatrix} = \begin{bmatrix} 0 & 1 & \vdots & 0 \\ -k/m & 0 & \vdots & 0 \\ \hline 0 & 0 & \vdots & 0 \end{bmatrix} \begin{bmatrix} x_c \\ v_c \\ \hline d_o \end{bmatrix} + \begin{bmatrix} 0 \\ \frac{1}{m} \\ \hline 0 \end{bmatrix} f$$

(4.33)

$$y = \begin{bmatrix} 1 & 0 & \vdots & -1 \end{bmatrix} \begin{bmatrix} x_c \\ v_c \\ \hline d_o \end{bmatrix}$$

(4.34)

Let the coefficient matrix of the first term on the right-hand side of Equation (4.33) be $\bar{A}$ and the coefficient matrix of Equation (4.34) be $\bar{C}$. Then, we obtain the observability matrix $U_o = \begin{bmatrix} \bar{C}^T & (\bar{C}\bar{A})^T & (\bar{C}\bar{A}^2)^T \end{bmatrix}^T$ and calculate the determinant of

$$det(U_o) = \begin{vmatrix} 1 & 0 & -1 \\ 0 & 1 & 0 \\ -\frac{k}{m} & 0 & 0 \end{vmatrix} = -\frac{k}{m}$$

Since $k \neq 0$ and $det(U_o) \neq 0$ is valid, the rank is 3, which is observable. It is unobservable if $k = 0$, $det(U_o) = 0$, and the rank is less than 3. So, the DOB is represented by Equations (4.35) and (4.36):

$$\begin{bmatrix} \dot{\hat{x}}_c \\ \dot{\hat{v}}_c \\ \dot{\hat{d}}_o \end{bmatrix} = \begin{bmatrix} 0 & 1 & 0 \\ -k/m & 0 & 0 \\ 0 & 0 & 0 \end{bmatrix} \begin{bmatrix} \hat{x}_c \\ \hat{v}_c \\ \hat{d}_o \end{bmatrix} + \begin{bmatrix} 0 \\ \frac{1}{m} \\ 0 \end{bmatrix} f + \begin{bmatrix} H_{11} \\ H_{12} \\ H_2 \end{bmatrix} (y - \hat{y})$$

$$(4.35)$$

$$\hat{y} = \begin{bmatrix} 1 & 0 & -1 \end{bmatrix} \begin{bmatrix} \hat{x}_c \\ \hat{v}_c \\ \hat{d}_o \end{bmatrix}$$

$$(4.36)$$

An example of a simulation is shown below. The use of output DOBs seems to be limited, given that the cart used is a common mechatronics system. Examples of the same dimensional output DOB are shown in List 4.5, and the Simulink model and waveforms are shown in Figures 4.13 and 4.14:

List 4.5: Example of the same dimensional output disturbance observer.

```
1   %% Physical Parameters
2   m=2; % Mass[kg]
3   k=3; % modulus of elasticity[N/m]
4   dis_step_xc=0.1;
5   %% State space model
6   A=[0,1;-k/m,0];B=[0;1/m];C=[1 0];% A,B,C matrices
7
8   %% Create extended system and output disturbance design
9   Abar=[A,zeros(2,1);zeros(1,2),0];
10  Bbar=[B;0];Cbar=[C,-1];
11  Ov=[Cbar;Cbar*Abar;Cbar*Abar^2];rank_of_Ov=rank(Ov) % Confirmation of rank
12  Q_ob=diag([1e1,1e1,1e5]);R_ob=1; % Q and R design of output DOB
13  Htemp=lqr(Abar',Cbar',Q_ob,R_ob);
14  H=Htemp';Ho1=H(1:2,:);Ho2=H(3,:);
15
16  %% State feedback controller design
17  pole=[-1,-1.5];F=place(A,B,pole);
18
19  %% Simulation
20  x0=[1.0;0.0];tend=10; % Initial values and simulation time
21  open('sim_Figure_4_13_and_4_14_Output_DOB');
22  set_param('sim_Figure_4_13_and_4_14_Output_DOB','WideLines','on');
23  set_param('sim_Figure_4_13_and_4_14_Output_DOB','ShowLineDimensions','on');
24  z=sim('sim_Figure_4_13_and_4_14_Output_DOB');
```

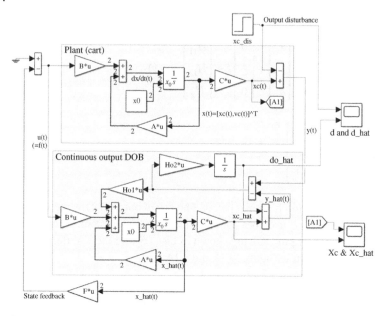

Figure 4.13 Example of a Simulink model of an output disturbance observer.

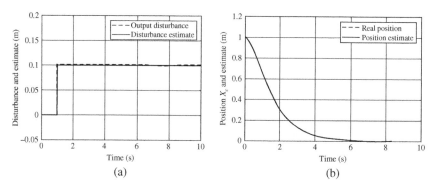

Figure 4.14 Simulation example of output disturbance observer. (a) Disturbances and disturbance estimate and (b) position and position estimate.

In Figure 4.13, the signal do_hat denotes the output disturbance $\hat{d}_o$. In Figure 4.14a, the disturbance estimation in the figure was fast but slightly oscillatory. The position waveform in Figure 4.14b is stable and convergent.[5]

---

[5] The LQ control method was used in this design example because the pole assignment method design did not provide a good response.

4.4 Identity Higher Order Disturbance Observer Design

The condition for a step disturbance (0-order disturbance) was $\dot{d}(t) = 0$. The ramp disturbance (first-order disturbance) represented by a first-order polynomial is defined by $\ddot{d}(t) = 0$. For example, if $d(t) = at + b$, then $\ddot{d}(t) = 0$. If the disturbance is represented by the second-degree polynomial (second-order disturbance)$d(t) = at^2 + bt + c$, then $\frac{d^3 d(t)}{dt^3} = 0$. Generalizing this, we consider the disturbance $d(t)$ to be a $p - 1$-order system ($p > 1$), which can be expressed by Equations (4.37) and (4.38), and collectively referred to as **higher order disturbances**.

$$\dot{\eta}(t) = \begin{bmatrix} \dot{\eta}_0 \\ \dot{\eta}_1 \\ \vdots \\ \dot{\eta}_{p-1} \end{bmatrix} = \Gamma \eta(t) = \Gamma \begin{bmatrix} \eta_0 \\ \eta_1 \\ \vdots \\ \eta_{p-1} \end{bmatrix}, \quad \Gamma = \begin{bmatrix} 0 & 1 & 0 & \cdots & 0 \\ 0 & 0 & 1 & \cdots & 0 \\ \vdots & \vdots & & \ddots & \vdots \\ 0 & 0 & 0 & \cdots & 1 \\ 0 & 0 & 0 & \cdots & 0 \end{bmatrix}$$

$$\Gamma \in R^{p \times p} \tag{4.37}$$

$$d(t) = H\eta(t), H = \begin{bmatrix} 1 & 0 & \cdots & 0 \end{bmatrix} \in R^{1 \times p} \tag{4.38}$$

The control plant is expressed as Equation (4.39):

$$\dot{x}(t) = Ax(t) + Bu(t) - Bd(t) = Ax(t) + Bu(t) - BH\eta(t) \tag{4.39}$$

Create Equations (4.40) and (4.41) with the state variables as $\bar{x}(t) = \begin{bmatrix} x^T(t) & \eta^T(t) \end{bmatrix}^T$.

$$\dot{\bar{x}}(t) = \begin{bmatrix} \dot{x}(t) \\ \dot{\eta}(t) \end{bmatrix} = \bar{A}\bar{x}(t) + \bar{B}u(t)$$

$$= \begin{bmatrix} A & -BH \\ 0 & \Gamma \end{bmatrix} \begin{bmatrix} x(t) \\ \eta(t) \end{bmatrix} + \begin{bmatrix} B \\ 0 \end{bmatrix} u(t) \tag{4.40}$$

$$y(t) = \bar{C}\bar{x}(t) = \begin{bmatrix} C & 0 \end{bmatrix} \begin{bmatrix} x(t) \\ \eta(t) \end{bmatrix} \tag{4.41}$$

If $(\bar{C}, \bar{A})$ is observable, then a $p - 1$-order DOB can be designed, which can be expressed as Equations (4.42)–(4.44):

$$\dot{\hat{\bar{x}}}(t) = \begin{bmatrix} \dot{\hat{x}}(t) \\ \dot{\hat{\eta}}(t) \end{bmatrix} = \bar{A}\hat{\bar{x}}(t) + \bar{B}u(t) + \hat{H}(y(t) - \hat{y}(t))$$

$$= \begin{bmatrix} A & -BH \\ 0 & \Gamma \end{bmatrix} \begin{bmatrix} \hat{x}(t) \\ \hat{\eta}(t) \end{bmatrix} + \begin{bmatrix} B \\ 0 \end{bmatrix} u(t) + \begin{bmatrix} H_1 \\ H_2 \end{bmatrix} (y(t) - \hat{y}(t)) \tag{4.42}$$

$$\hat{y}(t) = \bar{C}\hat{\bar{x}}(t) = \begin{bmatrix} C & 0 \end{bmatrix} \begin{bmatrix} \hat{x}(t) \\ \hat{\eta}(t) \end{bmatrix} \tag{4.43}$$

$$\hat{d}(t) = H\hat{\eta}(t) \tag{4.44}$$

Example 4.4.1 Consider a first-order DOB ($p = 1$) for a cart. Given that the order of the system is $n = 2$, the number of control inputs is $m = 1$, and the number of outputs is $l = 1$, and using $H = [1, 0]$, $\Gamma = \begin{bmatrix} 0 & 1 \\ 0 & 0 \end{bmatrix}$, we have List 4.6 and Figure 4.15.

List 4.6: Example of identity higher order disturbance observer.

```
1   %% Physical parameters
2   m=1.0;c=2.0; % Mass 2.0 kg, viscous friction coefficient 0.10 N/(m/s)
3   A=[0 1; 0 −c/m];B=[0;1/m];C=[1 0];
4   [n,m]=size(B); %Order and number of inputs
5
6   %% Controller design
7   pole=[−2.0,−3.0];F=place(A,B,pole);% State feedback gain design with pole assignment method
8
9   %% p−dimensional DOB design and setup for evaluation
10  p=1; % Designer decides dimension of disturbance DOB p=0 to 4 is observable
11  tend=5; % Setting the simulation time
12  dis_step=1.0; % Setting the magnitude of the time derivative of the stepwise disturbance
13
14  %% Setting up & expanding system for p−dimensional DOB design and evaluation
15  N=B;% Assume matching conditions are satisfied
16  H=1;Gamma=zeros(p+1,p+1);for i=1:p H=[H,0];end;
17  Gamma(1:p,2:p+1)=eye(p);% Create Gamma
18  Abar=[A,−N*H;zeros(p+1,n),Gamma];Bbar=[B;zeros(p+1,1)];
19  Cbar=[C,zeros(1,p+1)];
20
21  %% p−dimensional DOB design
22  pole_ob=[];base_of_pole=−15;delta_pole=2;
23  for i=1:n+p+1 pole_ob=[pole_ob,base_of_pole−delta_pole*i]; end;
24  Htemp=place(Abar',Cbar',pole_ob);Hob=Htemp';% Design gain by pole assignment method
25  H1=Hob(1:n);H2=Hob(n+1:n+p+1);% Estimated gain of state variables H1 and gain of disturbance
26  % estimation H2
27
28  %% Simulation
29  open('sim_Figure_4_15_and_4_16_higher_order_DOB');
30  x0=[0.05;0]; % Initial state value
31  set_param('sim_Figure_4_15_and_4_16_higher_order_DOB','WideLines','on');
32  set_param('sim_Figure_4_15_and_4_16_higher_order_DOB','ShowLineDimensions','on')
33  z=sim('sim_Figure_4_15_and_4_16_higher_order_DOB');
```

The waveforms of Figure 4.16a,b accurately estimate the ramp disturbance. However, when the initial value of the control plant is assumed to be unknown and non-0-valued, and the simulation is run with the initial value of the observer set to 0, the disturbance estimate converges after outputting an excessive

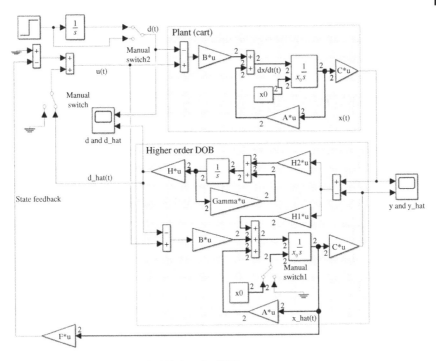

Figure 4.15 Simulink model of high-order DOB.

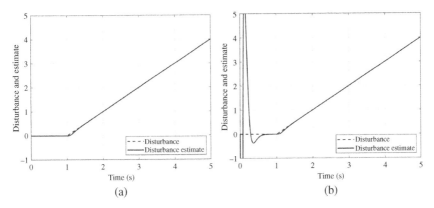

(a)   (b)

Figure 4.16 Disturbance and estimated value for ramped (first-order) disturbance observer. (a) With known initial disturbance and (b) with unknown initial disturbance.

impulse-like estimate. This phenomenon can also occur, so we need to be careful about whether to increase the disturbances' order or not.[6]

4.5 Minimal Order Disturbance Observer

Introducing the DOB design method with a **minimal order observer.**[7]

---

1. Create a matrix T of the form $T = \begin{bmatrix} C \\ W \end{bmatrix}$ such that $det(T) \neq 0$.

2. $TAT^{-1} = \begin{bmatrix} A_{11} & A_{12} \\ A_{21} & A_{22} \end{bmatrix}$, $TB = \begin{bmatrix} B_1 \\ B_2 \end{bmatrix}$.

   However, $A_{11} \in R^{l \times l}$, $A_{12} \in R^{l \times (n-l)}$, $A_{21} \in R^{(n-l) \times l}$, $A_{22} \in R^{(n-l) \times (n-l)}$, $B_1 \in R^{l \times m}$, and $B_2 \in R^{(n-l) \times m}$.

3. Determine $L \in R^{(n-l) \times l}$ where the real part of the eigenvalue of $\hat{A} = A_{22} - LA_{12}$ is negative.

4. $\hat{B} = \hat{A}L + A_{21} - LA_{11}$, $\hat{J} = -LB_1 + B_2$, $\hat{C} = T^{-1} \begin{bmatrix} 0_{l \times (n-l)} \\ cause\ I_{n-l} \end{bmatrix}$, $\hat{D} = T^{-1} \begin{bmatrix} I_l \\ L \end{bmatrix}$.

If we apply these to Figure 4.17, we can construct a minimal order observer. Let us design a simple minimal order DOB using the cart model.

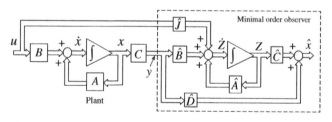

Figure 4.17 Block diagram of minimal order observer.

Figure 4.18 Minimal order disturbance observer with velocity observation.

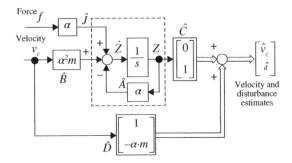

Example 4.5.1 (Simplest minimal order disturbance observer with velocity observation.)

(1) Basic Design Process

If the velocity v_c can be observed by a sensor (or estimated with reasonable accuracy), a simple structure **minimum-dimensional disturbance observer** can be constructed. Let $x = [x_1, x_2]^T = [v_c, d]^T$, $u = f$, and $y = v_c$. Using the equation of motion $m\dot{v}_c(t) = f(t) - d(t)$ and assuming $\dot{d}(t) = 0$, the equation of state is expressed as Equation (4.45), and the output equation as Equation (4.46):

$$\begin{bmatrix} \dot{v}_c \\ \dot{d} \end{bmatrix} = \begin{bmatrix} 0 & -1/m \\ 0 & 0 \end{bmatrix} \begin{bmatrix} v_c \\ d \end{bmatrix} + \begin{bmatrix} 1/m \\ 0 \end{bmatrix} f \tag{4.45}$$

$$y = \begin{bmatrix} 1 & 0 \end{bmatrix} \begin{bmatrix} v_c \\ d \end{bmatrix} \tag{4.46}$$

Following the above, let $W = \begin{bmatrix} 0 & 1 \end{bmatrix}$ and find $\hat{A}, \hat{B}, \hat{C}, \hat{D}, \hat{J}$. Then, we get $\hat{A} = A_{22} - LA_{12} = L/m = -\alpha$, $\hat{B} = \hat{A}L + A_{21} - LA_{11} = L^2/m = \alpha^2 m$, $\hat{C} = \begin{bmatrix} 0 \\ 1 \end{bmatrix}$, and $\hat{D} = \begin{bmatrix} 1 \end{bmatrix} = \begin{bmatrix} 1 \\ -\alpha J \end{bmatrix}$, where $\hat{J} = -LB_1 + B_2 = -L/m = \alpha$.

However, the observer's characteristic equation is $det(sI - \hat{A}) = s - L/m$, The observer's characteristic equation $det(sI - \hat{A}) = s - L/m$ is habitually written as $= s + \alpha$ and rewritten as $\alpha = -L/m$, i.e. $L = -\alpha m$.

Figure 4.18 represents it graphically.

Supplementary information for Figure 4.18.
(1) The first row of the output represents $\hat{v}_c$. It consists of the first row of $\hat{C}(= 0)z$ plus the first row of $\hat{D}(= 1)v_c$, which is just the observed output v_c itself.

(continued)

(*Continued*)

(2) The second row of the output represents $\hat{d}$. It consists of the second row of $\hat{C}(=1)z$ plus $-\alpha \cdot mv_c$. Meanwhile, the input side of the integrator, $\dot{z} = \alpha z + \alpha^2 mv_c + \alpha f$.

It can be organized by Laplace transforming to $z(s) = \frac{\alpha}{s+\alpha}\{m\alpha v_c(s) + f(s)\}$.

Adding up $-\alpha m v_c$ and organizing it,

we get $\hat{d}(s) = \frac{\alpha}{s+\alpha}\{f(s) + m\alpha v_c(s) - (s+\alpha)mv_c\} = \frac{\alpha}{s+\alpha}\{f(s) - msv_c(s)\}$.

Given that $d(s) = f(s) - msv_c(s)$, then $\hat{d}(s) = \frac{\alpha}{s+\alpha}d(s) = \frac{1}{Ts+1}d(s)$.

(2) Conversion to Transfer Function Representation

If we take out only the disturbance estimation part of this and organize it as a **transfer function** instead of the equation of state, we get the dashed line of Figure 4.19a, which can be further organized to get Figure 4.19b.

This representation form is simple and can be implemented in an operational amplifier circuit. Furthermore, if we take out the $\alpha/(s + \alpha)$ part of figure (b) and replace it with $T = 1/\alpha$, we get the original Figure 2.3.

Example 4.5.2 (Minimal order disturbance observer with position detection.) When the sensor can only observe the position x_c, a minimal order DOB with

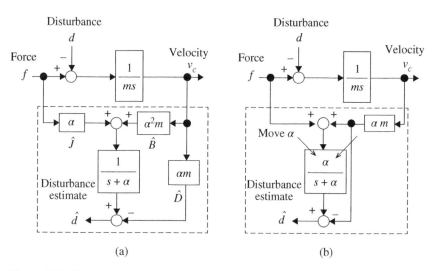

(a)   (b)

Figure 4.19 Equivalent disturbance observer with velocity observation. (a) Disturbance estimation part and (b) more organized representation (again).

a simple structure may be constructed. Unlike the previous example of velocity observation, let $x_1 = x_c$, $x_2 = v_c$, $x_3 = d$, $u = f$, and $y = x_1 = x_c$. In this case, the dimension is $n = 3$ with $p = 1$ outputs. Combined with the equation of motion $m\dot{v}_c = f - d$ and $\dot{d} = 0$, expanding system's equation is expressed in Equation (4.47).

$$\begin{bmatrix} \dot{x}_c \\ \dot{v}_c \\ \dot{d} \end{bmatrix} = \begin{bmatrix} 0 & 1 & 0 \\ 0 & 0 & -1/m \\ 0 & 0 & 0 \end{bmatrix} \begin{bmatrix} x_c \\ v_c \\ d \end{bmatrix} + \begin{bmatrix} 0 \\ 1/m \\ 0 \end{bmatrix} u \tag{4.47}$$

$$y = \begin{bmatrix} 1 & 0 & 0 \end{bmatrix} \begin{bmatrix} x_c \\ v_c \\ d \end{bmatrix} \tag{4.48}$$

$T = [C^T, W^T]^T = I.$ $W = \begin{bmatrix} 0 & 1 & 0 \\ 0 & 0 & 1 \end{bmatrix}$, and then let the gain matrix be $L = \begin{bmatrix} l_1 \\ l_2 \end{bmatrix}$, and $\hat{A} = A_{22} - LA_{12}$, $\hat{B} = \hat{A}L + A_{21} - LA_{11}$, $\hat{C} = T^{-1} \begin{bmatrix} 0_{p \times (n-p)} \\ cause\ I_{n-p} \end{bmatrix}$, and $\hat{D} = T^{-1} [I_p]$, respectively.

$$\hat{A} = \begin{bmatrix} 0 & -1/m \\ 0 & 0 \end{bmatrix} - \begin{bmatrix} l_1 \\ l_2 \end{bmatrix} \begin{bmatrix} 1 & 0 \end{bmatrix} = \begin{bmatrix} -l_1 & -1/m \\ -l_2 & 0 \end{bmatrix}$$

$$\hat{B} = \begin{bmatrix} -l_1 & -1/m \\ -l_2 & 0 \end{bmatrix} \begin{bmatrix} l_1 \\ l_2 \end{bmatrix} + \begin{bmatrix} 0 \\ 0 \end{bmatrix} - \begin{bmatrix} l_1 \\ l_2 \end{bmatrix} 0 = \begin{bmatrix} -l_1^2 - l_2/m \\ -l_1 l_2 \end{bmatrix}$$

$$\hat{C} = \begin{bmatrix} 0 & 0 \\ 1 & 0 \\ 0 & 1 \end{bmatrix}, \hat{D} = \begin{bmatrix} 1 \\ l_1 \\ l_2 \end{bmatrix}, \hat{J} = \begin{bmatrix} 1/m \\ 0 \end{bmatrix}$$

However, the characteristic equation of the observer $det(sI - \hat{A}) = s^2 + sl_1 - l_2/m$ is for clarity, and it is customary to write $s^2 + sg_1 + g_2$. If we rewrite it as $g_1 = l_1$ and $g_2 = -l_2/m$, the matrices are $\hat{A} = \begin{bmatrix} -g_1 & -1/m \\ g_2 m & 0 \end{bmatrix}$, $\hat{B} = \begin{bmatrix} -g_1^2 + g_2 \\ g_1 g_2 m \end{bmatrix}$, $\hat{D} = \begin{bmatrix} 1 \\ g_1 \\ -g_2 m \end{bmatrix}$.

Figure 4.20 represents this graphically.

Since $\hat{v}_c$ and $\hat{d}$ are exactly estimates, this observer can also be called "disturbance and velocity estimation observer" instead of "disturbance observer" [1, 2].

If we remove only the disturbance estimation function and organize it, we arrive at the dashed line in Figure 4.21a,b; Figure 4.21b seems to be the most commonly implemented configuration. To organize further, we go through Figure 4.22a and obtain Figure 4.22b as the final result. The value $\bar{e}$ is the calculated value of the deviation $e = f - d$ between the input force f and the disturbance d, which may include the observation noise, and the calculation error is not shown in the figure.

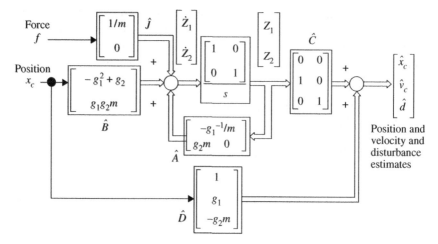

Figure 4.20 Position-observing disturbance observer.

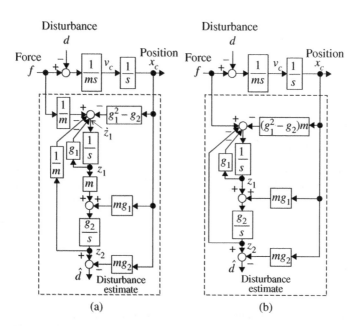

Figure 4.21 Equivalent disturbance observer with position observation. (a) Disturbance estimation part and (b) more organized representation.

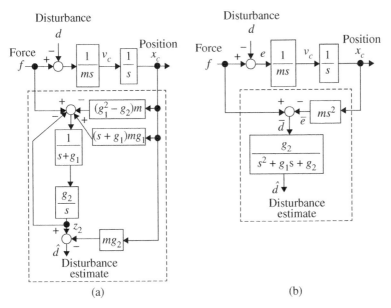

Figure 4.22 Equivalent disturbance observer conversion result with position observation. (a) Equivalent transformation 1 and (b) equivalent transformation 2.

Since $d = f - e$ for $e = f - d$, $\bar{d} = f - \bar{e}$ is obtained, and the disturbance estimate $\hat{d}$ is obtained through a second-order low-pass filter. For a mechatronics system with only a position sensor, a velocity-observing DOB or a DOB with position observation can be applied by calculating the derivative of the observed position information to obtain the velocity information.

Example 4.5.3 (Ramp (first-order) disturbance minimal order observer.) Let us introduce the minimal order observer for the ramp disturbance (first order) with velocity observation among the higher order disturbances represented by Equation (4.37). The state equation is Equation (4.49), and the output equation is Equation (4.50):

$$\begin{bmatrix} \dot{v}_c(t) \\ \dot{d}(t) \\ \ddot{d}(t) \end{bmatrix} = \begin{bmatrix} 0 & -\frac{1}{m} & 0 \\ 0 & 0 & 1 \\ 0 & 0 & 0 \end{bmatrix} \begin{bmatrix} v_c(t) \\ d(t) \\ \dot{d}(t) \end{bmatrix} + \begin{bmatrix} \frac{1}{m} \\ 0 \\ 0 \end{bmatrix} f(t) \tag{4.49}$$

$$y = \begin{bmatrix} 1 & 0 & 0 \end{bmatrix} \begin{bmatrix} v_c(t) & d(t) & \dot{d}(t) \end{bmatrix}^T \tag{4.50}$$

Design a minimal order observer using the procedure described above.

$$\hat{A} = A_{22} - LA_{12} = \begin{bmatrix} L_1/m & 1 \\ L_2/m & 0 \end{bmatrix} \quad \text{and} \quad det(\lambda I - \hat{A}) = \lambda^2 - \frac{L_1}{m} - \frac{L_2}{m} = \lambda^2 + g_1\lambda + g_2 =$$

$(\lambda - p_1)(\lambda - p_2)$, here the p1 and p2 equivalent to the poles of the observer. Using the polynomial coefficients g_1 and g_2 to denote each coefficient matrix, $\hat{A}$ becomes

$$\hat{A} = \begin{bmatrix} -g_1 & 1 \\ -g_2 & 0 \end{bmatrix} \quad \text{and} \quad \hat{B} = \begin{bmatrix} m(g_1^2 - g_2) \\ mg_1g_2 \end{bmatrix}, \quad \hat{C} = \begin{bmatrix} 0 & 0 \\ 1 & 0 \\ 0 & 1 \end{bmatrix}, \quad \hat{D} = \begin{bmatrix} 1 \\ -mg_1 \\ -mg_2 \end{bmatrix}, \quad \text{and} \quad \hat{J} = \begin{bmatrix} g_1 \\ g_2 \end{bmatrix}.$$

Example programs can be found in List 4.7, Simulink examples in Figure 4.23, and the disturbance waveform is shown in Figure 4.24.

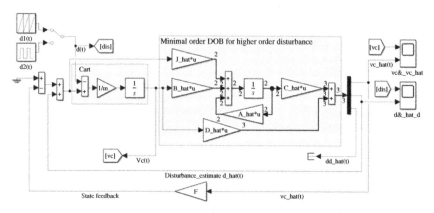

Figure 4.23 Simulink model of DOB for ramp (first-order) disturbance.

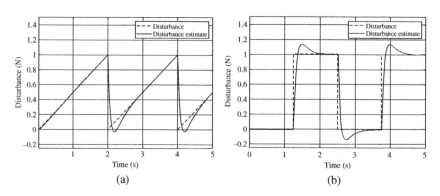

Figure 4.24 Waveform of ramp (first-order) disturbance observer. (a) Estimate for triangular disturbance and (b) estimate for step disturbance.

List 4.7: Example of ramp (first-order) disturbance minimal order observer with velocity observation.

```
1   %% Physical parameters
2   m=2.0; % Mass[kg]
3
4   %% Expanded system state space model
5   A=0;B=1/m;C=1;D=0;
6   A_bar=[A -B,0;zeros(2,2),[1;0]];
7   A11=A_bar(1,1);A12=A_bar(1,2:3);A21=A_bar(2:3,1);A22=A_bar(2:3,2:3);
8   B_bar=[B;zeros(2,1)];B1=B_bar(1);B2=B_bar(2:3);
9   C_bar=[C,zeros(1,2)];D_bar=0;
10
11  %% first-order disturbance minimal order observer design
12  p1=-8;p2=-9; % Observer poles and polynomial coefficients
13  g1=-(p1+p2);g2=p1*p2; % and polynomial coefficients
14  % Coefficient of characteristic equation of A_hat 's^2+g1s+g2'
15  L1=-m*g1;L2=-m*g2;L=[L1;L2];
16  A_hat=A22-L*A12;B_hat=A_hat*L+A21-L*A11;
17  C_hat=[zeros(1,2);eye(2)];D_hat=[1;L];J_hat=-L*B1+B2;
18
19  %% Control system design with LQ control method
20  Q=1.5;R=2;F=lqr(A,B,Q,R);
21
22  %% Simulation
23  vc0=0;d=1.0;t_end=5.0;
24  open_system('sim_Figure_4_23_and_4_24_min_1order_DOB');
25  set_param('sim_Figure_4_23_and_4_24_min_1order_DOB','WideLines','on');
26  set_param('sim_Figure_4_23_and_4_24_min_1order_DOB','ShowLineDimensions','on')
27  y=sim('sim_Figure_4_23_and_4_24_min_1order_DOB');
```

As the name suggests, the first-order DOB can successfully estimate the triangular wave-like disturbance corresponding to the ramp disturbance(first-order) disturbance, as shown in Figure 4.24a; Figure 4.24b shows that we can estimate the step disturbance, but there is an overshoot.

4.6 Design of Periodic Disturbance Observer

We consider a design method where the disturbance is represented by a sinusoidal wave **periodic disturbance** [3, 4]. In practice, this book focuses on an observer that estimates the basic sinusoidal disturbance.

Consider a periodic disturbance in a single-input system. Let us assume that the disturbance $d(t) = a_{ex} \sin \omega_{ex} t$ is a sinusoidal wave of angular frequency ω_{ex}

and amplitude a_{ex}. Its transfer function is $G_d(s) = \frac{a_{ex}\omega_{ex}}{s^2+\omega_{ex}^2}$, and the disturbance is generated by $d(s) = G_d(s)\cdot\mathcal{L}[\delta(t)]$.[8]

If this is the state variable $x_d = [x_{d1}, x_{d2}]^T$, the disturbance model can be expressed, for example, as a controllable canonical system, as in Equations (4.51) and (4.52):

$$\begin{bmatrix} \dot{x}_{d1}(t) \\ \dot{x}_{d2}(t) \end{bmatrix} = \begin{bmatrix} 0 & 1 \\ -\omega_{ex}^2 & 0 \end{bmatrix} \begin{bmatrix} x_{d1}(t) \\ x_{d2}(t) \end{bmatrix} + \begin{bmatrix} 0 \\ 1 \end{bmatrix} \delta(t) \tag{4.51}$$

$$d(t) = \begin{bmatrix} a_{ex}\omega_{ex} & 0 \end{bmatrix} \begin{bmatrix} x_{d1}(t) \\ x_{d2}(t) \end{bmatrix} \tag{4.52}$$

Let the coefficient matrices on the right-hand side of Equation (4.51) be $A_w \in R^{2\times2}$ and $B_w \in R^{2\times1}$ and the coefficient matrix of Equation (4.52) be $C_w \in R^{1\times2}$. When dealing with a periodic disturbance model with multiple inputs and frequencies, the number of inputs, outputs, and orders should be determined according to the model. However, for the control target represented by $\dot{x}(t) = Ax(t) + Bu(t) - Bd(t)$ and $y(t) = Cx(t)$ in the m-input l-output n-order system, an extended system is created, yielding Equations (4.53) and (4.54):

$$\begin{bmatrix} \dot{x}(t) \\ \dot{x}_d(t) \end{bmatrix} = \begin{bmatrix} A & -BC_w \\ 0 & A_w \end{bmatrix} \begin{bmatrix} x(t) \\ x_d(t) \end{bmatrix} + \begin{bmatrix} B \\ 0 \end{bmatrix} u(t) + \begin{bmatrix} 0 \\ B_w \end{bmatrix} \delta(t) \tag{4.53}$$

$$y = \begin{bmatrix} C & 0 \end{bmatrix} \begin{bmatrix} x(t) \\ x_d(t) \end{bmatrix} \tag{4.54}$$

One feature is that the upper right element of the $\bar{A}$ matrix of the expanded system is $-BC_w$, not $-B$. Next, when observability holds, we create Equations (4.55)–(4.57) as identity observers:

$$\begin{bmatrix} \dot{\hat{x}}(t) \\ \dot{\hat{x}}_d(t) \end{bmatrix} = \begin{bmatrix} A & -BC_w \\ 0 & A_w \end{bmatrix} \begin{bmatrix} \hat{x}(t) \\ \hat{x}_d(t) \end{bmatrix} + \begin{bmatrix} B \\ 0 \end{bmatrix} u(t) + \begin{bmatrix} H_1 \\ H_2 \end{bmatrix} (y - \hat{y}) \tag{4.55}$$

$$\hat{y} = \hat{x}_c = \overline{C}\hat{x} = \begin{bmatrix} C & 0 \end{bmatrix} \begin{bmatrix} \hat{x}(t) \\ \hat{x}_d(t) \end{bmatrix} \tag{4.56}$$

$$\hat{d} = \overline{C}_s\hat{x} = \begin{bmatrix} 0, C_w \end{bmatrix} \begin{bmatrix} \hat{x}(t) \\ \hat{x}_d(t) \end{bmatrix} \tag{4.57}$$

The observed output and estimated equation are in the form $y = \overline{C}x$ as before, but the disturbance estimate $\hat{d}(t)$ is in a different equation (4.57) than the observed output equation. Additionally, the design method for the observer gain is the same as before, using the pole assignment and LQ control methods.

---

8 sin function is represented by $\frac{a_{ex}}{s^2+\omega^2}$. The subscript *ex* refers to "external." An impulse response is required for a sinusoidal wave to be generated.

Example 4.6.1 (Control system with periodic disturbance observer for the cart system.) The simple equations of motion of the cart assuming the position observation are as follows: $m\frac{dv_c(t)}{dt} = f(t) - d(t)$, and the coefficient matrices of the equation of state with $x = [x_c, v_c]^T$ are $A = \begin{bmatrix} 0 & 1 \\ 0 & 0 \end{bmatrix}$, $B = \begin{bmatrix} 0 \\ \frac{1}{m} \end{bmatrix}$, and $C = \begin{bmatrix} 1 & 0 \end{bmatrix}$.

The observer using the enlarged system has Equations (4.58)–(4.60). Incorporating the periodic disturbance model as an internal model in the expanded system is important. We consider that $H_2(y - \hat{y})$ generates the disturbance estimate instead of the impulse input.

$$
\begin{bmatrix} \dot{\hat{x}}_c \\ \dot{\hat{v}}_c \\ \dot{\hat{x}}_{d1} \\ \dot{\hat{x}}_{d2} \end{bmatrix} = \begin{bmatrix} 0 & 1 & 0 & 0 \\ 0 & 0 & -\frac{a_{ex}\omega_{ex}}{m} & 0 \\ 0 & 0 & 0 & 1 \\ 0 & 0 & -\omega_{ex}^2 & 0 \end{bmatrix} \begin{bmatrix} \hat{x}_c \\ \hat{v}_c \\ \hat{x}_{d1} \\ \hat{x}_{d2} \end{bmatrix} + \begin{bmatrix} 0 \\ \frac{1}{m} \\ 0 \\ 0 \end{bmatrix} f
$$

$$
+ \begin{bmatrix} H_{11} & H_{12} & H_{21} & H_{22} \end{bmatrix}^T (y - \hat{y}) \tag{4.58}
$$

$$
\hat{y} = \hat{x}_c = \begin{bmatrix} 1 & 0 & 0 & 0 \end{bmatrix} \begin{bmatrix} \hat{x}_c & \hat{v}_c & \hat{x}_{d1} & \hat{x}_{d2} \end{bmatrix}^T \tag{4.59}
$$

$$
\hat{d} = \begin{bmatrix} 0 & 0 & a_{ex}\omega_{ec} & 0 \end{bmatrix} \begin{bmatrix} \hat{x}_c & \hat{v}_c & \hat{x}_{d1} & \hat{x}_{d2} \end{bmatrix}^T \tag{4.60}
$$

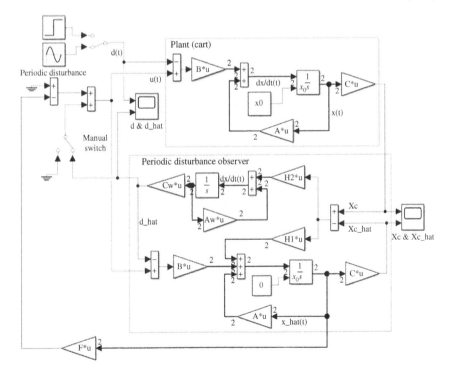

Figure 4.25 Simulink model of periodic disturbance observer.

An example program is shown in List 4.8; Figure 4.25 shows the Simulink model; and the simulation results are shown in Figure 4.26.

It can be seen that the periodic disturbance estimate, $\hat{d}$, in Figure 4.26a-1 perfectly follows the specified periodic disturbance d. Figure 4.26b-1 shows that the position waveform also converges to zero after a transient delay.

However, Figure 4.26a-2 shows that the step-like disturbance cannot be estimated, as it should be, and (b-2) shows a steady deviation.

What happens when a periodic disturbance twice as large as the estimated period is added? As shown in Figure 4.26a-3, an estimation error occurs, and the vibration at the position shown in (b-3) remains. Meanwhile, when we examined the case where the amplitude of the periodic disturbance changes, we found that we could estimate the amplitude without any problem. For reasons of space, the figure is omitted.

List 4.8: Periodic disturbance observer (position observed).

```
1   %% Physical Parameters
2   m=2; % Mass[kg]
3   dis_step=1; % Magnitude of disturbance
4   w=2*pi*1;a=1; % Angular frequency w and magnitude a
5   % of sinusoidal disturbance
6   Aw=[0,1;-w^2,0];Bw=[0;1];Cw=[a*w,0];
7   wex=w*1;aex=a*1; % Angular frequency and amplitude of real disturbance.
8   % Not necessarily equal.
9
10  %% State space model
11  A=[0,1;0,0];B=[0;1/m];C=[1 0];% ABC matrix of continuous system
12
13  %% Creating an extended system and designing the observer gain
14  Abar=[A,-B*Cw;zeros(2,2),Aw];
15  Bbar=[B;zeros(2,1)];Cbar=[C,zeros(1,2)];
16  pole_ob=[-60,-65,-70,-75]*0.2; % With pole assignment method
17  Htemp=place(Abar',Cbar',pole_ob);H=Htemp';
18  H1=H(1:2);H2=H(3:4); % H1 is for state est., H2 is for disturbance est.
19
20  %% Design of state feedback gain
21  pole=[-3,-4];F=place(A,B,pole);
22
23  %% Simulation
24  x0=[0.0];tend=5; % Initial values and simulation time
25  amp_pulse=1;Ts=10;puls_width=10/amp_pulse;
26  open_system('sim_Figure_4_25_and_4_26_periodic_dob');
27  set_param('sim_Figure_4_25_and_4_26_periodic_dob','WideLines','on');
28  set_param('sim_Figure_4_25_and_4_26_periodic_dob','ShowLineDimensions','on');
29  z=sim('sim_Figure_4_25_and_4_26_periodic_dob');
```

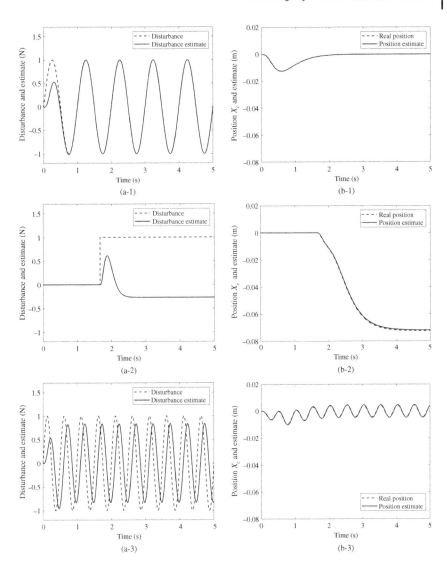

Figure 4.26 Simulation example of periodic disturbance observer. (a-1) Disturbance and estimate (periodic disturbance), (a-2) disturbance and estimate (step disturbance), (a-3) disturbance and estimate (periodic disturbance, twice frequency), (b-1) position and estimate (periodic disturbance), (b-2) position and estimate (step disturbance), and (b-3) position and estimate (periodic disturbance, twice frequency).

4.7 Observability and Noninput/Output Disturbances

To design an observer, i.e. estimate, a state variable, the system must be **observable**.[9] In this section, we discuss the observability condition in the context of a **DC motor**.

4.7.1 Mathematical Model of a DC Motor

In Section 2.7, we introduced three equations to describe the motion of a DC motor. Replacing the rotation equation (2.27) with the equation of motion yields the following:

$$\tau(t) = J\frac{d\omega(t)}{dt} + k\theta(t) + \tau_d(t) \tag{4.61}$$

The $k\theta$ added to the new equation (4.61) is an elastic torque assuming a torsional spring that generates a torque proportional to the rotation angle. Although torsional springs and the like are not installed in general motor drive systems, they were added for learning **observability**. Moreover, no current feedback is provided in this section. An image of the object is shown in Figure 4.27.

In the above equation, the initial values of the rotation angle, rotation velocity, and drive current are zero. The initial values of the rotation angle, rotation velocity, and drive current are Laplace transformed to zero and represented in the block diagram as Figure 4.28.[10]

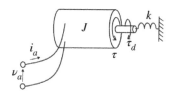

Figure 4.27 Image of a DC motor with a torsion spring.

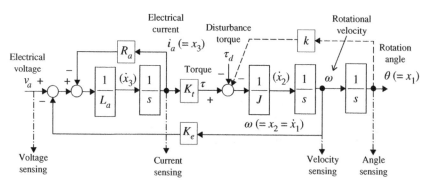

Figure 4.28 Transfer function block diagram of DC motor.

---

9 Appendix A.3.4.
10 In general, most systems have $k = 0$, so the signal lines before and after are dashed.

Note that the disturbance τ_d in Figure 4.28 is neither an input nor an output as described previously.[11] In this book, we refer to them as **noninput/output disturbances**.

4.7.2 DC Motor Observable Matrix and Rank

Since we need the state and output equations of the DC motor model to investigate the observability, we define the state variables following the terms in Appendix A.3.1. As shown in Figure 4.28, $x = \begin{bmatrix} x_1 & x_2 & x_3 \end{bmatrix}^T = \begin{bmatrix} \theta & \omega & i_a \end{bmatrix}^T$, with control input $u = v_a$ and disturbance $d = \tau_d$. For the observed output, assume that $y = \theta$, $y = \omega$ when a rotation angle sensor, rotation velocity sensor, and current sensor, respectively, are assumed to be present. Let us assume that only one type of sensor is installed. From Figure 4.28, $\dot{x}_1(t) = \dot{\theta}(t) = \omega(t)$, and $\dot{x}_2(t) = \dot{\omega}(t) = \frac{1}{J}(K_t i_a(t) - \tau_d(t) - k\theta(t))$. Furthermore, we get $\dot{x}_3(t) = \dot{i}_a(t) = \frac{1}{L_a}(v_a(t) - K_e\omega(t) - R_a i_a(t))$. Summarizing $\dot{x} = Ax + Bu - B_{dis}d$ and $y = Cx$, we get the following equation:

$$
\begin{bmatrix} \dot{\theta}(t) \\ \dot{\omega}(t) \\ \dot{i}_a(t) \end{bmatrix} = \begin{bmatrix} 0 & 1 & 0 \\ -\frac{k}{J} & 0 & \frac{K_t}{J} \\ 0 & -\frac{K_e}{L_a} & -\frac{R_a}{L_a} \end{bmatrix} \begin{bmatrix} \theta(t) \\ \omega(t) \\ i_a(t) \end{bmatrix} + \begin{bmatrix} 0 \\ 0 \\ \frac{1}{L_a} \end{bmatrix} v_a(t)
$$
$$
- \begin{bmatrix} 0 & \frac{1}{J} & 0 \end{bmatrix}^T \tau_d(t) \tag{4.62}
$$

$$
y(t) = \begin{cases} C_1 x = [1, 0, 0]x & \text{(when a rotation angle sensor is installed)} \\ C_2 x = [0, 1, 0]x & \text{(when a rotational velocity sensor is installed)} \\ C_3 x = [0, 0, 1]x & \text{(when an electrical current sensor is installed)} \end{cases} \tag{4.63}
$$

Before examining the observability, let us first calculate the observability matrix assuming that there is no disturbance $B_{dis} = 0$. Since the order is $n = 3$ and the number of outputs is $l = 1$, we calculate the observability matrix according to Appendix A.3.5. The observability matrix is observable if **rank** is equal to the order n, and unobservable if rank is less than the order n.[12] If there is a rotation angle sensor, the following equation can be obtained:

$$
U_{o1} = \begin{bmatrix} C_1 \\ C_1 A \\ C_1 A^2 \end{bmatrix} = \begin{bmatrix} 1 & 0 & 0 \\ 0 & 1 & 0 \\ -k/J & 0 & K_t/J \end{bmatrix}
$$

Next, $det(U_{o1}) = K_t/J$ holds.[13] In other words, $rank(U_{o1}) = 3$ regardless of the value of k, which is observable.

---

11 As shown in Section 2.7, if current feedback is used or L_a can be ignored, it is considered an input disturbance. If L_a is negligible, it can be treated as an input disturbance.
12 For rank, see Appendix A.1.3.
13 For a one-output system, if $det(U_o) \neq 0$, the rank is consistent with order n, so it is observable.

If a rotational velocity sensor is present, the following equation can be obtained:

$$U_{o2} = \begin{bmatrix} C_2 \\ C_2 A \\ C_2 A^2 \end{bmatrix} = \begin{bmatrix} 0 & 1 & 0 \\ -k/J & 0 & K_t/J \\ 0 & -\frac{k}{J} - \frac{K_e K_t}{J L_a} & -\frac{K_t R_a}{J L_a} \end{bmatrix}$$

where $det(U_{o2}) = \frac{kK_t R_a}{J^2 L_a}$, and $det(U_{o2}) = 0$ for $k = 0$.

If we calculate the rank again

$$rank(U_{o2}) = \begin{cases} 3 & \text{observable (when } k \neq 0) \\ 2 & \text{unobservable (when } k = 0) \end{cases}$$

If a current sensor is present,

$$U_{o3} = \begin{bmatrix} C_3 \\ C_3 A \\ C_3 A^2 \end{bmatrix} = \begin{bmatrix} 0 & 0 & 1 \\ 0 & -K_e/L_a & -R_a/L_a \\ \frac{K_e k}{J L_a} & \frac{K_e R_a}{L_a^2} & \frac{R_a^2}{L_a^2} - \frac{K_e K_t}{J L_a} \end{bmatrix}$$

Then, we get the following: $det(U_{o3}) = \frac{kK_e^2}{J L_a^2}$. If $k = 0$, then the following holds: $det(U_{o3}) = 0$.

As in the previous example, we calculate the rank,

$$rank(U_{o3}) = \begin{cases} 3 & \text{observable (when } k \neq 0) \\ 2 & \text{unobservable (when } k = 0) \end{cases}$$

What does this mean? In summary, we obtain Table 4.1.

Generally, since we do not have torsion springs, etc., we know that $k = 0$, and the rotation angle cannot be estimated from the rotational velocity information.[14]

In Figure 4.28, when $k = 0$, there is no path (=dashed line) where the angle θ information is transmitted to the "velocity observation" and "current observation" parts.

Table 4.1 Observability comparison 1.

Sensor	Observability	Estimable physical quantity
Angle sensor	Observable	θ, ω, i_a
Velocity sensor	Observable if there is a spring	θ, ω, i_a
Current sensor	Observable if there is a spring	θ, ω, i_a

---

14 When the initial value of the angle is known, for a short period, the angle can be found by integrating the velocity information over time, but the error accumulates as time passes.

Table 4.2 Observability comparison 2.

Sensor	Observability	Estimable physical quantity
Angle sensor	Observable	$\theta, \omega, i_a, \tau_d$
Velocity sensor	Unobservable	Whether there is a spring or not, nothing can be estimated
Current sensor	Unobservable	Whether there is a spring or not, nothing can be estimated

However, when the torsion spring of the mechatronics part is mounted, $k \neq 0$, the angle information is transmitted from the signal line of the single-dotted line to the velocity or current observation part. In particular, when the current value is observed, it is possible to design an angle sensorless control system in which the mechatronics physical quantity (angle θ, velocity ω) is estimated from the electrical physical quantity (voltage v_a, current i_a) alone [5, 6].

4.7.3 Observability of Disturbance Estimation

What about the estimation of τ_d, which is a non-input–output disturbance? Assume that the disturbance $\dot{d} = \dot{\tau}_d = 0$, and the state variables are $\bar{x} = [\theta, \omega, i_a, \tau_d]^T$. The order is $n + 1 = 4$. Deriving the state and output equations using the expanded state variables, creating the observability matrix $\bar{U}_o$, and examining the rank, we obtain the following conclusion.

If there is a rotation angle sensor, then $det(\bar{U}_{o1}) = -\frac{K_t R_a}{J^2 L_a}$. It is full rank at $rank(\bar{U}_{o1}) = 4$, regardless of the value of k, so it is observable. If there is a rotational velocity sensor, $det(\bar{U}_{o2}) = 0$, it is not observable. If there is a current sensor, $det(\bar{U}_{o3}) = 0$, not observable. In summary, we can have Table 4.2.

Let us examine Figure 4.28. This is because the disturbance τ_d and the torsional reaction force of the spring $k\theta$ are added at the same point and cannot be distinguished from the velocity and current observation parts. Thus, observers or DOBs cannot be always designed for any control plant [5, 6].

4.7.4 Noninput/Output Disturbance Observer and Control

Let us try to estimate the actual noninput/output disturbance, τ_d. Let us add the program corresponding to Figure 4.3 to List 4.9. The Simulink model is shown in Figure 4.29.

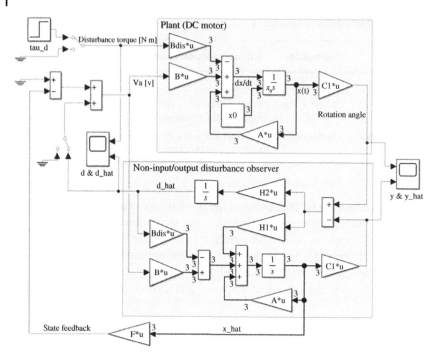

Figure 4.29 Simulink model of DOB for DC motors with large inductance.

List 4.9: Noninput/output disturbance estimation for DC motors.

```
1   %% Physical parameters
2   Ra=0.65;La=5e-2; % Armature resistance & inductance
3   Ke=2.8e-2; % Inductive electromotive force constant
4   Kt=Ke; % Torque constant
5   J=1e-1; % Rotor moment of inertia
6   k=0; % Modulus of elasticity of torsion spring
7   % attached to motor. The value of k depends on the system
8
9   %% State space model
10  % State variable x=[theta(t);omega(t);i(t)], u=va(t)
11  A=[0, 1, 0;
12      -k/J, 0, Kt/J;
13      0, -Ke/La, -Ra/La];
14  B=[0;0;1/La];Bdis=[0;1/J;0];C1=[1 0 0];D=0;
15
16  %% Create extended system and observer design
17  A_=[A -Bdis;zeros(1,4)];B_=[B;0];C1_=[C1 0];D_=D;
```

```
18   pole_ob=[-5,-6,-7,-8];H_temp=place(A_',C1_',pole_ob);
19   H=H_temp';H1=H(1:3);H2=H(4);
20
21   %% Stabilized control system design
22   pole_fb=[-1.2,-1.3,-1.4];F=place(A,B,pole_fb);
23
24   open('sim_Figure_4_29_and_4_30_for_dc_motordc_motor');
25   t_end=20;x0=[0;0;0];tau_d=1; % tau_d is the magnitude of disturbance torque
26   set_param('sim_Figure_4_29_and_4_30_for_dc_motordc_motor','WideLines','on');
27   set_param('sim_Figure_4_29_and_4_30_for_dc_motordc_motor','ShowLineDimensions','on');
28   z=sim('sim_Figure_4_29_and_4_30_for_dc_motordc_motor');
```

It should be noted here that there is a difference between the B matrix and the B_{dis} matrix. Although the adder connects them in the second stage of both blocks, they do not satisfy the so-called matching condition, because the former has nonzero components only in the third row, whereas the latter has nonzero components only in the second row.

In Figure 4.29, we dare to positively feedback the disturbance torque estimate $\hat{\tau}_d$. Although the disturbance can be estimated in Figure 4.30a, the control input is the drive voltage v_a, and the disturbance torque τ_d cannot be canceled directly. Therefore, it can be read from Figure 4.30b that the rotation angle cannot converge to zero.[15]

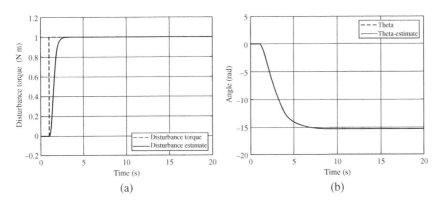

(a) (b)

Figure 4.30 Simulation results of DOB for DC motors with large inductance. (a) Disturbance and estimate and (b) angle and estimate.

---

15 The advantage of this observer is that the state can be estimated considering the disturbance, although it cannot cancel the disturbance.

References

1 Akira Shimada, Tsuyoshi Umeda, Norio Yokoshima, Naoki Kawawada, Hiroshi Watanabe, Toshimi Shioda, Masahide Nagai: Motion Control for Vertically Articulated Direct Drive Robot Manipulator, IEEE International Workshop on Advanced Motion Control (AMC'90), 132–137, 1990.

2 Yuhki Kosaka, Akira Shimada: Motion Control for Articulated Robots Based on Accurate Modeling, The 8th IEEE International Workshop on Advanced Motion Control (AMC'04), 535–540, 2004.

3 Hisayoshi Muramatsu, Seiichiro Katsura: An adaptive periodic-disturbance observer for periodic-disturbance suppression, IEEE Transactions on Industrial Electronics, Vol. 14, No. 10, 4446–4456, 2018.

4 Hiroshi Fujimoto, Fumihiro Kawakami, Seiji Kondo: Multirate Repetitive Control and Applications - Verification of Switching Scheme by HDD and Visual Servoing, Proceedings of the 2003 American Control Conference, 2875–2880, 2003.

5 Akira Shimada, Kohji Enomoto: Realization of position sensor-less control system using spring, IEEJ Transactions of Industrial Application, Vol. 124, No. 12, 1268–1273, 2004 (In Japanese).

6 Akira Shimada, Yu Kishiwada, Michiyo Arimura: AC servo motor position sensorless control using mechanical springs, IEEE 9th International Workshop on Advanced Motion Control (AMC'06), 559–562, 2006.

5

Digital Disturbance Observer Design

Disturbance observer design based on digital control theory is essential and valuable in actual mechatronics systems, considering the effects of control cycles.

5.1 Identity Digital Disturbance Observer Design

From Equations (4.3) and (4.4), the control plant is represented by the state and output equations in Equations (5.1) and (5.2):

$$\dot{x}(t) = Ax(t) + Bu(t) - Bd(t) \tag{5.1}$$
$$y(t) = Cx(t) \tag{5.2}$$

Assuming the control period T and the input with the **zero-order hold** function, it converts the plant to the digital system, yielding Equations (5.3) and (5.4). We obtain the discrete system equations of state and output equations. Also, the disturbance property $\dot{d} = 0_{m\times1}$ can be replaced by $d(k+1) = d(k) \in R^m$:

$$x(k+1) = A_d x(k) + B_d u(k) - B_d d(k) \tag{5.3}$$
$$y(k) = C_d x(k) \tag{5.4}$$

We create the extended systems (5.5) and (5.6), putting these relations together. That is, since $\bar{x}(k) = [x(k)^T, d(k)^T]^T$, we obtain

$$\bar{x}(k+1) = \bar{A}_d \bar{x}(k) + \bar{B}_d u(k) \tag{5.5}$$
$$y(k) = \bar{C}_d \bar{x}(k) \tag{5.6}$$

However,

$$\bar{A}_d = \begin{bmatrix} A_d & -B_d \\ 0 & I \end{bmatrix}, \ \bar{B}_d = \begin{bmatrix} B_d \\ 0 \end{bmatrix}, \text{ and } \bar{C}_d = \begin{bmatrix} C_d & 0 \end{bmatrix}$$

Disturbance Observer for Advanced Motion Control with MATLAB/Simulink, First Edition. Akira Shimada.
© 2023 The Institute of Electrical and Electronics Engineers, Inc. Published 2023 by John Wiley & Sons, Inc.
Companion website: www.wiley.com/go/disturbanceobserver

If the extended system is observable, then we can design an **identity digital disturbance observer** [1, 2]; $\overline{H}_d = [H_1^T, H_2^T]^T$ is the gain matrix of the observer.

$$\hat{\bar{x}}(k+1) = \bar{A}_d\hat{\bar{x}}(k) + \bar{B}_d u(k) + \overline{H}_d\{y(k) - \hat{y}(k)\} \tag{5.7}$$

$$\hat{y}(k) = \overline{C}_d\hat{\bar{x}}(k) \tag{5.8}$$

The specific description of the content is as follows:

$$\begin{bmatrix} \hat{x}(k+1) \\ \hline \hat{d}(k+1) \end{bmatrix} = \begin{bmatrix} A_d & \vline & -B_d \\ \hline 0 & \vline & I \end{bmatrix} \cdot \begin{bmatrix} \hat{x}(k) \\ \hline \hat{d}(k) \end{bmatrix}$$
$$+ \begin{bmatrix} B_d \\ \hline 0 \end{bmatrix} u(k) + \begin{bmatrix} H_1 \\ \hline H_2 \end{bmatrix} \{y(k) - \hat{y}(k)\} \tag{5.9}$$

$$\hat{y}(k) = \begin{bmatrix} C_d & \vline & 0 \end{bmatrix} \cdot \begin{bmatrix} \hat{x}(k) \\ \hline \hat{d}(k) \end{bmatrix} \tag{5.10}$$

Necessarily, the coefficient matrix should be denoted as $\bar{A}_d$, $\bar{B}_d$, $\overline{H}_d$, and $\overline{C}_d$. Figure 5.1 is this observer's block diagram. Picking up each element from this observer, we acquire Equations (5.11), (5.12), and (5.13). Then, we can draw Figure 5.2 combining them. Taking out each element of this observer, we get Equation (5.11):

$$\hat{x}(k+1) = A_d\hat{x}(k) + B_d u(k) - B_d \cdot \hat{d}(k)$$
$$+ H_1(y(k) - \hat{y}(k)) \tag{5.11}$$

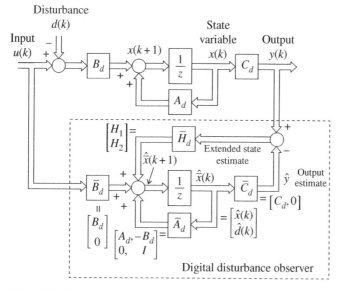

Figure 5.1 Block diagram of digital disturbance observer (DOB).

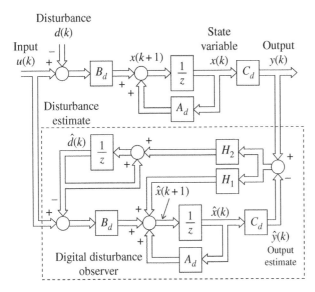

Figure 5.2 Block diagram of a digital DOB (individual element representation).

$$\hat{d}(k+1) = \hat{d}(k) + H_2(y(k) - \hat{y}(k)) \tag{5.12}$$

$$\hat{y}(k) = C_d\hat{x}(k) \tag{5.13}$$

Now, if the observer gain $\overline{H}_d = [H_1, H_2]$, its design is chosen so that the absolute value of the root of $\bar{A}_d - \overline{H}_d\overline{C}_d$, which is the digital pole, is less than 1.

Example 5.1.1 (Identity digital disturbance observer with velocity observation.) Design a digital disturbance observer for a cart system with velocity observation. Let $x[k] = [x_1(k), x_2(k)]^T = [v_c(k), d(k)]^T$, $u(k) = f(k)$, and $y(k) = v_c(k)$, and assume that $d(k+1) = d(k)$. If we assume that $d(k+1) = d(k)$, then we have the state and output equations of the digital system corresponding to Equations (4.45) and (4.46) in Equations (5.14) and (5.15):[1]

$$\begin{bmatrix} v_c(k+1) \\ d(k+1) \end{bmatrix} = \begin{bmatrix} 1 & -T/m \\ 0 & 1 \end{bmatrix} \begin{bmatrix} v_c(k) \\ d(k) \end{bmatrix} + \begin{bmatrix} T/m \\ 0 \end{bmatrix} f(k) \tag{5.14}$$

$$y(k) = \begin{bmatrix} 1 & 0 \end{bmatrix} \begin{bmatrix} v_c(k) \\ d(k) \end{bmatrix} \tag{5.15}$$

---

1 In MATLAB [Ad,Bd,Cd,Dd]=c2dm(A,B,C,D,T), but the same can be obtained using the approximation formula and **forward difference**. That is, from $m\frac{v_c(k+1)-v_c(k)}{T} = f(k) - d(k)$, we get $v_c(k+1) = v_c(k) + \frac{T}{m}f(k) - \frac{T}{m}d(k)$. $d(k+1) = d(k)$ corresponds to $\dot{d}(t) = 0$.

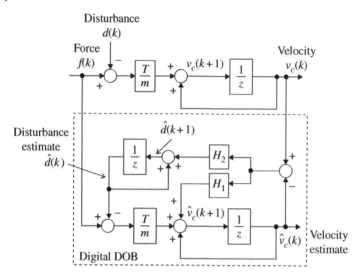

Figure 5.3 Block diagram of digital identity DOB with velocity observation.

The block diagram then becomes Figure 5.3, and the identity digital disturbance observer is represented by Equations (5.16) and (5.17):

$$\begin{bmatrix} \hat{v}_c(k+1) \\ \hat{d}(k+1) \end{bmatrix} = \begin{bmatrix} 1 & -T/m \\ 0 & 1 \end{bmatrix} \begin{bmatrix} \hat{v}_c(k) \\ \hat{d}(k) \end{bmatrix} + \begin{bmatrix} T/m \\ 0 \end{bmatrix} f(k)$$

$$+ \begin{bmatrix} H_1 \\ H_2 \end{bmatrix} \{v_c(k) - \hat{v}_c(k)\} \tag{5.16}$$

$$\hat{y}(k) = \begin{bmatrix} 1 & 0 \end{bmatrix} \begin{bmatrix} \hat{v}_c(k) \\ \hat{d}(k) \end{bmatrix} \tag{5.17}$$

The observer gain is chosen so that the absolute value of all eigenvalues λ_i ($i = 1, 2$) of $\bar{A}_d - \overline{H}\,\overline{C}_d$ is less than 1. So, for example, if $\lambda_1 = p_1$ and $\lambda_2 = p_2$, then

$$\left| \begin{bmatrix} \lambda & 0 \\ 0 & \lambda \end{bmatrix} - \left(\begin{bmatrix} 1 & -T/m \\ 0 & 1 \end{bmatrix} - \begin{bmatrix} H_1 \\ H_2 \end{bmatrix} \begin{bmatrix} 1 & 0 \end{bmatrix} \right) \right|$$

$$= \lambda^2 + (H_1 - 2)\lambda + (1 - H_1 - H_2 T/m)$$

$$= (\lambda - p_1)(\lambda - p_2) = \lambda^2 - (p_1 + p_2)\lambda + p_1 p_2 \tag{5.18}$$

From this, we get $H_1 = -(p_1 + p_2) + 2$ and $H_2 = m/T \cdot (1 - H_1 - p_1 p_2)$.

The corresponding example programs are shown in List 5.1, and the model diagrams and waveforms using Simulink are shown in Figures 5.4 and 5.5.

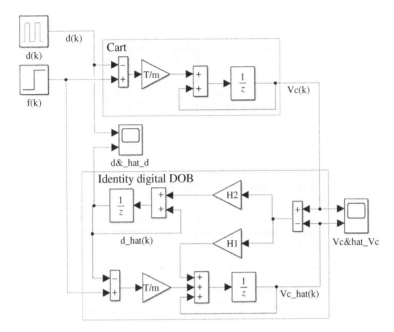

Figure 5.4 Simulink model of digital identity DOB with velocity observation.

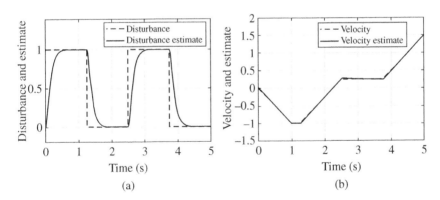

Figure 5.5 Simulation result digital identity DOB with velocity observation. (a) Disturbance and estimate and (b) velocity and estimate.

List 5.1:   Example of identity digital DOB with velocity observation.

```
1   %% Physical parameters
2   m=1.0; % Mass[kg]
3
4   %% Extended system state-space model
5   A=[0 -1/m;0,0];B=[1/m;0];C=[1,0];D=0;
```

```
6   [n,~]=size(A);[~,m]=size(B);[p,~]=size(C);
7   T=0.01;[Ad,Bd,Cd,Dd]=c2dm(A,B,C,D,T);
8   % or Ad=[0,−T/m;0,1],Bd=[T/m;0]
9
10  %% Disturbance observer design
11  pole_ob=[0.87,0.92];
12  H=place(Ad',Cd',pole_ob);H1=H(1);H2=H(2);
13
14  %% Simulation
15  f=0.5;d=1.0;t_end=5.0;
16  open_system('sim_Figure_5_4_and_5_5_identity_digital_dob_with_vel');
17  y=sim('sim_Figure_5_4_and_5_5_identity_digital_dob_with_vel');
```

In addition to the estimation of the disturbance in Figure 5.5a, it can be observed that the velocity estimate's waveform is superimposed on the velocity waveform in (b).

Example 5.1.2 (Identity digital disturbance observer with position observation.) Design a digital disturbance observer assuming a cart with positional observation. Here, the state variable $x[k] = [x_c(k), v_c(k)]^T$, the control input $u(k) = f(k)$, and the observation output $y(k) = x_c(k)$. The state and output equations are the equations of a continuous system. These equations are discretized using the continuous system equations (4.10) and (4.11) and are expressed as follows:[2]

$$\begin{bmatrix} x_c(k+1) \\ v_c(k+1) \end{bmatrix} = \begin{bmatrix} 1 & T \\ 0 & 1 \end{bmatrix} \begin{bmatrix} x_c(k) \\ v_c(k) \end{bmatrix} + \begin{bmatrix} \frac{1}{2}T^2/m \\ T/m \end{bmatrix} f(k)$$
$$- \begin{bmatrix} \frac{1}{2}T^2/m \\ T/m \end{bmatrix} d(k) \tag{5.19}$$

$$y(k) = \begin{bmatrix} 1 & 0 \end{bmatrix} \begin{bmatrix} x_c(k) \\ v_c(k) \end{bmatrix} \tag{5.20}$$

Next, let us have the expanded system as $d(k+1) = d(k), x[k] = [x_c(k), v_c(k), d(k)]^T$, $u(k) = f(k)$, and $y(k) = x_c(k)$.

$$\begin{bmatrix} x_c(k+1) \\ v_c(k+1) \\ d(k+1) \end{bmatrix} = \begin{bmatrix} 1 & T & -\frac{1}{2}T^2/m \\ 0 & 1 & -T/m \\ 0 & 0 & 1 \end{bmatrix} \begin{bmatrix} x_c(k) \\ v_c(k) \\ d(k) \end{bmatrix} + \begin{bmatrix} \frac{1}{2}T^2/m \\ T/m \\ 0 \end{bmatrix} f(k) \tag{5.21}$$

---

2 To acquire the coefficient matrices A_d, B_d, C_d, and Dd, we can use MATLAB's c2dm function easily. Or, we derive them by fitting $x(t) = x(0) + v(0)t + \frac{1}{2}at^2$, $ma = f - d$, and $x_c(k+1) = x_c(k) + Tv_c(k) + \frac{1}{2}\frac{T^2}{m}\{f(k) - d(k)\}$ and fitting vc(k + 1) similarly. If we simply put $\frac{x_c(k+1)-x_c(k)}{T} = v_c(k)$, we can get $x_c(k+1) = x_c(k) + Tv_c(k)$. However, we do not recommend it because it is different from the general digital system equation of state assuming a zero-order hold.

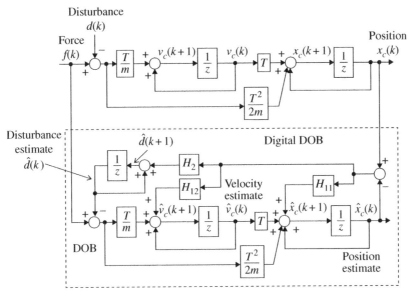

Figure 5.6 Block diagram of identity digital disturbance observer with position observation.

$$y(k) = \begin{bmatrix} 1 & 0 & | & 0 \end{bmatrix} \begin{bmatrix} x_c(k) \\ v_c(k) \\ \hline d(k) \end{bmatrix} \tag{5.22}$$

Here, each coefficient matrix is denoted by $\bar{A}_d$, $\bar{B}_d$, and $\bar{C}_d$, and each element separated by a dashed line is denoted by $\bar{A}_{d11} = A_d$, $\bar{A}_{d12} = -B_d$, $\bar{A}_{d21} = [0,0]$, $\bar{A}_{d22} = 1$, $\bar{B}_{d1} = [0, T/m]^T$, $\bar{B}_{d2} = 0$, $\bar{C}_{d1} = C_d$, and $\bar{C}_{d2} = 0$. Based on the above, the identity digital disturbance observer is represented as in Equations (5.23) and (5.24), and the block diagram is generally represented in Figure 5.1. However, it can also be represented element by element as shown in Figure 5.6.

$$\begin{bmatrix} \hat{x}_c(k+1) \\ \hat{v}_c(k+1) \\ \hline \hat{d}(k+1) \end{bmatrix} = \begin{bmatrix} 1 & T & | & -\frac{1}{2}T^2/m \\ 0 & 1 & | & -T/m \\ \hline 0 & 0 & | & 1 \end{bmatrix} \begin{bmatrix} \hat{x}_c(k) \\ \hat{v}_c(k) \\ \hline \hat{d}(k) \end{bmatrix} + \begin{bmatrix} \frac{1}{2}T^2/m \\ T/m \\ \hline 0 \end{bmatrix} f(k)$$

$$+ \begin{bmatrix} H_{11} \\ H_{12} \\ \hline H_2 \end{bmatrix} \{x_c(k) - \hat{x}_c(k)\} \tag{5.23}$$

$$\hat{y}(k) = \begin{bmatrix} 1 & 0 & | & 0 \end{bmatrix} \begin{bmatrix} \hat{x}_c \\ \hat{v}_c(k) \\ \hline \hat{d}(k) \end{bmatrix} \tag{5.24}$$

5.2 Confirmation of Separation Theorem

The control input cannot make the unknown disturbance any desired value because it is evident that the disturbance itself is unaffected by the control input. However, even knowing that the extended system with disturbance is uncontrollable, let us attempt the following state feedback control:

$$
\begin{aligned}
u(k) &= -F \cdot [\hat{x}(k)^T, \hat{d}(k)^T]^T \\
&= -[F_1, -F_2][\hat{x}^T(k), \hat{d}^T(k)]^T \\
&= -F_1\hat{x}(k) + F_2\hat{d}(k)
\end{aligned}
\tag{5.25}
$$

Substituting Equations (5.3) and (5.11) into Equation (5.25) and putting it together, we get Equation (5.26):

$$
\begin{bmatrix} x(k+1) \\ d(k+1) \\ \hat{x}(k+1) \\ \hat{d}(k+1) \end{bmatrix} = \begin{bmatrix} A_d & -B_d & -B_dF_1 & B_dF_2 \\ 0 & I & 0 & 0 \\ H_1C_d & 0 & \alpha_1 & \alpha_2 \\ H_2C_d & 0 & -H_2C_d & I \end{bmatrix} \begin{bmatrix} x(k) \\ d(k) \\ \hat{x}(k) \\ \hat{d}(k) \end{bmatrix}
\tag{5.26}
$$

where $\alpha_1 = A_d - B_dF_1 - H_1C_d$ and $\alpha_2 = -B_d + B_dF_2$.
Here, the error vector $\begin{bmatrix} e_x^T & e_d^T \end{bmatrix}^T$ is defined as follows:

$$
\begin{bmatrix} e_x(k) \\ e_d(k) \end{bmatrix} = \begin{bmatrix} \hat{x}(k) \\ \hat{d}(k) \end{bmatrix} - \begin{bmatrix} x(k) \\ d(k) \end{bmatrix}
\tag{5.27}
$$

Furthermore, assuming a new state variable $\begin{bmatrix} x^T & d^T & e_x^T & e_d^T \end{bmatrix}^T$ and applying a coordinate transformation to Equation (5.26) yields Equation (5.28):

$$
\begin{bmatrix} x(k+1) \\ d(k+1) \\ e_x(k+1) \\ e_d(k+1) \end{bmatrix} = \begin{bmatrix} \alpha_3 & \alpha_2 & -B_dF_1 & B_dF_2 \\ 0 & I & 0 & 0 \\ 0 & 0 & \alpha_4 & -B_d \\ 0 & 0 & -H_2C_d & I \end{bmatrix} \cdot \begin{bmatrix} x(k) \\ d(k) \\ e_x(k) \\ e_d(k) \end{bmatrix}
\tag{5.28}
$$

where $\alpha_3 = A_d - B_dF_1$ and $\alpha_4 = A_d - H_1C_d$.
Through coordinate transformation, the system's poles are not changed. Therefore, we can check the eigenvalues λ of this coefficient matrix to verify the poles of the control system.

$$
\begin{aligned}
&\det \begin{bmatrix} \lambda I - \alpha_3 & -\alpha_2 & B_dF_1 & -B_dF_2 \\ 0 & \lambda I - I & 0 & 0 \\ 0 & 0 & \lambda I - \alpha_4 & B_d \\ 0 & 0 & H_2C_d & \lambda I - I \end{bmatrix} \\
&= \begin{vmatrix} \lambda I - \alpha_3 & -\alpha_2 \\ 0 & \lambda I - I \end{vmatrix} \cdot \begin{vmatrix} \lambda I - \alpha_4 & B_d \\ H_2C_d & \lambda I - I \end{vmatrix} \\
&= |\lambda I - \alpha_3| \cdot |\lambda I - I| \cdot \begin{vmatrix} \lambda I - \alpha_4 & B_d \\ H_2C_d & \lambda I - I \end{vmatrix}
\end{aligned}
\tag{5.29}
$$

The right-hand side of Equation (5.29) consists of the product of three terms. The first term corresponds to the poles of the control system with the state feedback applied to the control target excluding the disturbance, and the second term corresponds to the m stability limit poles of the step disturbance, i.e. the origin poles of the continuous system. The third term is the pole of the disturbance observer in Equation (5.9). From the above, we can see the following two points:

1. The **separation theorem** holds for the state feedback system using an identity disturbance observer. In other words, the poles of the control system and the poles of the disturbance observer can be set separately.
2. The poles of the control system, including the observer, are independent of the disturbance estimate's feedback gain matrix F_2.

5.3 Minimal Order Digital Disturbance Observer

The design of the **minimal order digital observer** is based on the following procedure:

(1) Create a matrix T_d of the form $T_d = \begin{bmatrix} C_d \\ W_d \end{bmatrix}$ such that $det(T_d) \neq 0$.

(2) $T_d A_d T_d^{-1} = \begin{bmatrix} A_{d11} & A_{d12} \\ A_{d21} & A_{d22} \end{bmatrix}$ $T_d B_d = \begin{bmatrix} B_{d1} B_{d2} \end{bmatrix}$.

However, $A_{d11} \in R^{l \times l}$, $A_{d12} \in R^{p \times (n-l)}$, $A_{d21} \in R^{(n-l) \times p}$, $A_{d22} \in R^{(n-l) \times (n-l)}$, $B_{d1} \in R^{l \times m}$, $B_{d2} \in R^{(n-l) \times m}$.

(3) Determine $L_d \in R^{(n-l) \times l}$ so that the magnitude of the eigenvalues of $\hat{A}_d = A_{d22} - L_d A_{d12}$ is less than 1.

(4) $\hat{B}_d = \hat{A}_d L_d + A_{d21} - L_d A_{d11}$, $\hat{J}_d = -L_d B_{d1} + B_{d2}$, $\hat{C}_d = T_d^{-1} \begin{bmatrix} 0_{l \times (n-l)} \\ I_{n-l} \end{bmatrix}$,

$\hat{D}_d = T_d^{-1} \begin{bmatrix} I_l \\ L_d \end{bmatrix}$.

Figure 5.7 shows the block diagram of the minimal order digital observer corresponding to the above design process. Based on this, we design a minimal order digital disturbance observer by creating an extended system that includes the disturbance d. Let us illustrate an example with velocity observation.

Example 5.3.1 (Minimal order digital disturbance observer with velocity observation.) Design a minimal order disturbance observer assuming a cart with velocity observation as a digital system. As in the case of the same dimension, let $x[k] = [x_1(k), x_2(k)]^T = [v_c(k), d(k)]^T$, $u(k) = f(k)$, and $y(k) = v_c(k)$. Also, let $d(k + 1) = d(k)$. The state and output equations are Equations (5.14) and (5.15).[3]

_____

3 Let each coefficient matrix be A_d, B_d, and C_d, and the i, j components of each matrix be denoted as A_{dij}.

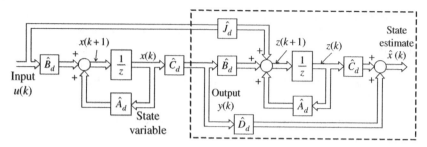

Figure 5.7 Block diagram of minimal order digital observer.

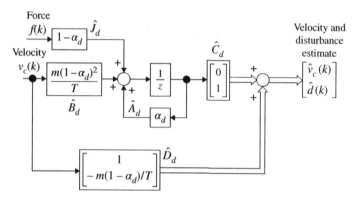

Figure 5.8 Minimal order digital disturbance observer in velocity observation.

Assuming that $W = \begin{bmatrix} 0 & 1 \end{bmatrix}$, using $S = \begin{bmatrix} C \end{bmatrix}$, since $S = I$. $SA_d S^{-1} = A_d$, $SB_d = B_d$, and $\hat{A}_d = A_{d22} - L_d A_{d12} = 1 - L_d(-T/m) = 1 + TL_d/m$, $\hat{B} = \hat{A}_d L_d + A_{d21} - L_d A_{d11} = L_d^2 T/m$, $\hat{C} = \begin{bmatrix} 0 \\ 1 \end{bmatrix}$, $\hat{D} = \begin{bmatrix} 1 \\ L_d \end{bmatrix}$, and $\hat{J} = -LT/m$.

However, the observer's characteristic equation is $det(zI - \hat{A}_d) = z - (1 + TL_d/m)$ and rewrite it as $\alpha_d = 1 + TL_d/m$, i.e. $L_d = -(1 - \alpha_d)m/T$. $\hat{A}_d = \alpha_d$, $\hat{B}_d = (1 - \alpha_d)^2 m/T$, $\hat{D}_d = \begin{bmatrix} 1 \\ -m(1 - \alpha_d)/T \end{bmatrix}$, and $\hat{J}_d = 1 - \alpha_d$. If we make a block diagram of this, we can draw Figure 5.8. As we did with the continuous system, let us extract only the disturbance estimation function and express it in the **transfer function**. Then, after (a) and (b) in Figure 5.9, we get (c). The $\bar{d}(k)$ is the computational value of the disturbance, and the $(1 - \alpha_d)/(z - \alpha_d)$ is the first-order digital low-pass filter, and the disturbance estimate $\hat{d}(k)$ is the output.

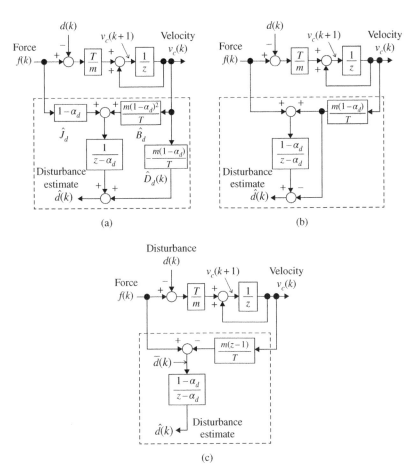

Figure 5.9 Transformation of minimal order digital DOB for velocity observation.
(a) Expression with disturbance estimation part, (b) more organized expression, and
(c) equivalent digital DOB with velocity observation.

Let us check the properties of $G(z) = \frac{1-\alpha_d}{z-\alpha_d}$ just to be sure. Consider the unit step input $u(z) = \frac{z}{z-1}$ and output $y(z) = G(z)u(z)$, under $|\alpha_d| < 1$. Using the final value theorem, we have

$$y(\infty) = \lim_{z \to 1} \left(\frac{z-1}{z} \right) y(z) = \lim_{z \to 1} \left(\frac{z-1}{z} \right) \left\{ \frac{1-\alpha_d}{z-\alpha_d} \frac{z}{z-1} \right\} = 1$$

Additionally, if we calculate the response as $T = 0.01$, $\alpha_d = exp(-aT)$, we get Figure 5.10, which is equivalent to the first-order delay of a continuous system. The corresponding MATLAB program is shown in List 5.2, and the Simulink model and disturbance estimation waveforms are shown in Figures 5.11 and 5.12.

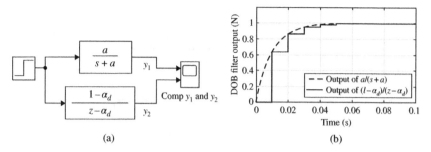

(a)   (b)

Figure 5.10 Comparison of $G(s) = \frac{a}{s+a}$ and $G(z) = \frac{1-\alpha_d}{z-\alpha_d}$. (a) Simulink model and (b) wave form.

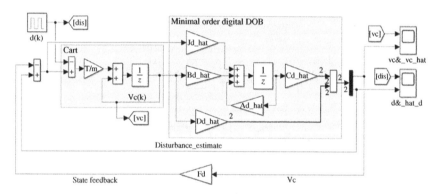

Figure 5.11 Simulink model of minimal order digital DOB with velocity observation.

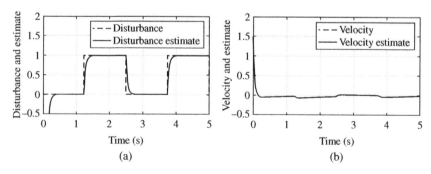

(a)   (b)

Figure 5.12 Simulation result of minimal order digital DOB with velocity observation. (a) Disturbance and estimation and (b) velocity and estimation.

List 5.2:   Minimal order digital disturbance observer with velocity observation.

```
1   %% Physical paremeters
2   m=1.0; % Mass[kg]
3
4   %% Extended system model
5   A=0;B=1/m;C=1;D=0;
6   Abar=[A −B;0,0];Bbar=[B;0];Cbar=[C,0];Dbar=0;
7   [n,~]=size(Abar);[~,m]=size(Bbar);[p,~]=size(Cbar);
8   T=0.01;[Adbar,Bdbar,Cdbar,Ddbar]=c2dm(Abar,Bbar,Cbar,Dbar,T);
9   % or Adbar=[1,−T/m;0,1],Bdbar=[T/m;0]
10
11  %% Disturbance observer design
12  alpha_d=0.8; % Pole of digital DOB
13  Ld=−(1−alpha_d)*m/T;
14  Ad_hat=Adbar(2,2)−Ld*Adbar(1,2); % Ad_hat=1+T*L/m=alpha
15  Bd_hat=Ad_hat*Ld+Adbar(2,1)−Ld*Adbar(1,1);
16  % Bd_hat=T*Ld^2/m=(1−alpha_d)^2*m/T
17  Cd_hat=[0;1];Dd_hat=[1;Ld]; % Dd_hat=[1;−(1−alpha_d)m/T]
18  Jd_hat=−Ld*Bdbar(1)+Bdbar(2); % J_hat=−Ld*T/m=1−alpha_d
19
20  %% Controller design
21  [Ad,Bd,Cd,Dd]=c2dm(A,B,C,D,T);
22  Q=1.5;R=2;Fd=dlqr(Ad,Bd,Q,R); % Design with LQ control method
23
24  %% Simulation
25  f=0.5;d=1.0;t_end=5.0;vc0=1.5;
26  open_system('sim_Figure_5_11_and_5_12_min_digital_DOB');
27  set_param('sim_Figure_5_11_and_5_12_min_digital_DOB','WideLines','on');
28  set_param('sim_Figure_5_11_and_5_12_min_digital_DOB','ShowLineDimensions','on');
29  y=sim('sim_Figure_5_11_and_5_12_min_digital_DOB');
```

The velocity feedback gain was obtained using the state feedback gain calcula-tion function dlqr.m for LQ[4] control of a discrete-time system in MATLAB.[5] The velocity estimation waveforms in Figure 5.12b are completely overlapped since they are only passed through the real velocity.

Example 5.3.2 (Minimal order digital disturbance observer with position obser-vation.)   To design a digital observer assuming a cart by position observation [19], the state and output equations of the extended system, which are the same as Equations (5.21) and (5.22), are used. Here, the division into the small matrices is different since the position information is the observable information, and the

---

4 See Appendix A.3.7 about LQ control.
5 LQ control is briefly described in Appendix A.3.

variables to be estimated are velocity and disturbance. Consider Equations (5.30) and (5.31):

$$
\begin{bmatrix} x_c(k+1) \\ \hline v_c(k+1) \\ d(k+1) \end{bmatrix} = \begin{bmatrix} 1 & T & -\frac{1}{2}T^2/m \\ \hline 0 & 1 & -T/m \\ 0 & 0 & 1 \end{bmatrix} \begin{bmatrix} x_c(k) \\ \hline v_c(k) \\ d(k) \end{bmatrix} + \begin{bmatrix} \frac{1}{2}T^2/m \\ T/m \\ 0 \end{bmatrix} f(k) \tag{5.30}
$$

$$
y(k) = \begin{bmatrix} 1 & 0 & 0 \end{bmatrix} \begin{bmatrix} x_c(k) \\ \hline v_c(k) \\ d(k) \end{bmatrix} \tag{5.31}
$$

We denote each coefficient matrix as A_d, B_d, and C_d, and each element separated by a dashed line as $A_{d11} = 1$, B_d, and C_d.

$A_{d12} = [T, -\frac{T^2}{2m}]$, $A_{d21} = \begin{bmatrix} 0 \\ 0 \end{bmatrix}$, $A_{d22} = \begin{bmatrix} 1 & -\frac{T}{m} \\ 0 & 1 \end{bmatrix}$, $B_{d1} = \frac{T^2}{2m}$, $B_{d2} = \begin{bmatrix} \frac{T}{m} \\ 0 \end{bmatrix}$, $C_{d1} = 1$, and $C_{d2} = \begin{bmatrix} 0 & 0 \end{bmatrix}$.

$W = \begin{bmatrix} 0 & 1 & 0 \\ 0 & 0 & 1 \end{bmatrix}$ as $S = \begin{bmatrix} C \\ W \end{bmatrix}$, since $S = I_3$.

Let $SA_dS^{-1} = A_d$ and $SB_d = B_d$, and let $L_d = [L_1, L_2]^T$.

$$
\hat{A}_d = A_{d22} - L_d A_{d12} = \begin{bmatrix} a_{11} & a_{12} \\ a_{21} & a_{22} \end{bmatrix} = \begin{bmatrix} 1 - L_1 T & \frac{L_1 T^2}{2m} - \frac{T}{m} \\ -L_2 T & \frac{L_2 T^2}{2m} + 1 \end{bmatrix}
$$

$$
\hat{B}_d = \hat{A}_d L_d + A_{d21} - L_d A_{d11} = \begin{bmatrix} b_1 \\ b_2 \end{bmatrix} = \begin{bmatrix} \frac{L_2 T(L_1 T - 2)}{2m} - L_1^2 T \\ \frac{L_2^2 T^2}{2m} - L_1 L_2 T \end{bmatrix}
$$

$$
\hat{J}_d = -L_d B_{d1} + B_{d2} = \begin{bmatrix} j_1 \\ j_2 \end{bmatrix} = \begin{bmatrix} -\frac{L_1 T^2 - 2T}{2m} \\ -\frac{L_2 T^2}{2m} \end{bmatrix}
$$

$\hat{C}_d = [0_{2\times1}, I_2]^T$, $\hat{D}_d = [1, L_d]^T$

Let $\hat{A}_d$ be the stabilizing roots p_1 and p_2, then from $det(\lambda I_2 - \hat{A}_d) = (\lambda - p_1)(\lambda - p_2) = 0$, we have

$$
L_d = \begin{bmatrix} L_1 & L_2 \end{bmatrix}^T = \begin{bmatrix} \frac{p_1 + p_2 - p_1 p_2 + 3}{2T} & -\frac{m(p_1+1)(p_2+1)}{T^2} \end{bmatrix}^T.
$$

If we make this a block diagram, we can draw Figure 5.13. An example program is shown in List 5.3, and the Simulink and estimated waveforms are shown in Figures 5.14 and 5.15. An example of the poles of the control system for obtaining L_d is the pair of $p_1 = 0.75$ and $p_2 = 0.80$.

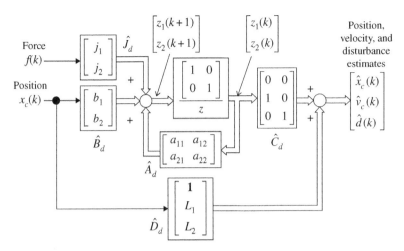

Figure 5.13 Minimal order digital disturbance observer in position observation.

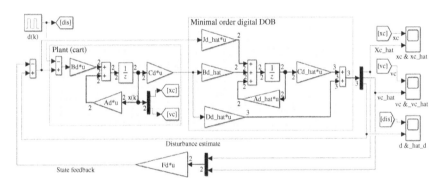

Figure 5.14 Simulink model of minimal order digital DOB in position observation.

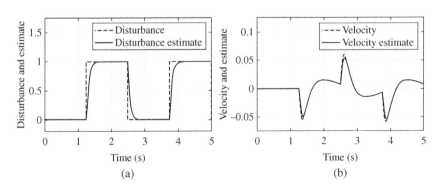

Figure 5.15 Simulation results for a minimal order digital DOB in position observation. (a) Disturbance and estimate and (b) velocity and estimate.

List 5.3: Minimal order digital DOB with position observation.

```
1    %% Physical paremeters
2    m=1;T=0.01; % Mass[kg] and control ratio[s]
3
4    %% Extended system model
5    A=[0 1;0 0];B=[0;1/m];C=[1 0];D=0;T=0.01;
6    [Ad,Bd,Cd,Dd]=c2dm(A,B,C,D,T);
7    A_=[A,-B;zeros(1,2),0];B_=[B;0];C_=[C,0];% 拡大系の作成
8    [Ad_,Bd_,Cd_,Dd_]=c2dm(A_,B_,C_,D,T);
9    % or Ad=[1,T,-1/2*T^2/m;0,1,-T/m;0,0,1];Bd=[1/2*T^2/m;T/m;0];Cd=[1, 0,0];
10   % Wd=[0,1,0;0,0,1];Td=[C_d;W_d]=eye(3);inv(Td)*Ad*Td=Ad
11   A11=Ad_(1,1);A12=Ad_(1,2:3);A21=Ad_(2:3,1);A22=Ad_(2:3,2:3);
12   B1=Bd_(1);B2=Bd_(2:3);
13
14   %% Disturbance observer design
15   pole=[0.75,0.80];Ltemp=place(A22',A12',pole);Ld=Ltemp';
16   Ad_hat=A22-Ld*A12;Bd_hat=Ad_hat*Ld+A21-Ld*A11;
17   Jd_hat=-Ld*B1+B2;Cd_hat=[zeros(1,2);eye(2)];Dd_hat=[1;Ld];
18
19   %% Controller desig
20   Q=diag([1e2,1e1]);R=1;Fd=dlqr(Ad,Bd,Q,R);
21
22   %% Simulation
23   t_end=5;d=1.0;x0=[0.0;0];
24   open_system("sim_Figure_5_14_and_5_15_min_dig_DOB_pos");
25   set_param('sim_Figure_5_14_and_5_15_min_dig_DOB_pos','WideLines','on');
26   set_param('sim_Figure_5_14_and_5_15_min_dig_DOB_pos','ShowLineDimensions','on');
27   y=sim('sim_Figure_5_14_and_5_15_min_dig_DOB_pos');
```

It can be seen that both disturbance and velocity estimations are well done.

Example 5.3.3 (Digital system disturbance observer and control system represented by transfer function.) Now that we have converted the minimum-dimensional digital disturbance observer to the transfer function, let us try digitally representing the velocity PI control system introduced in Section 2.3.3. The same cart model is used. The disturbance observer in Figure 5.16 is configured in the form of Figure 5.9b. The waveforms of Column (a) in Figure 5.17 are obtained using the positive feedback with the Manual switch ON for the disturbance estimate for the no-noise system. However, the waveforms of Column (b) are obtained using the Manual switch ON for the system with noise. The magnitude of the noise and PI controller's gain is the same, and the parameter α_d is the digital pole $\alpha_d = exp(-aT)$, equivalent to the pole $-a$ of the

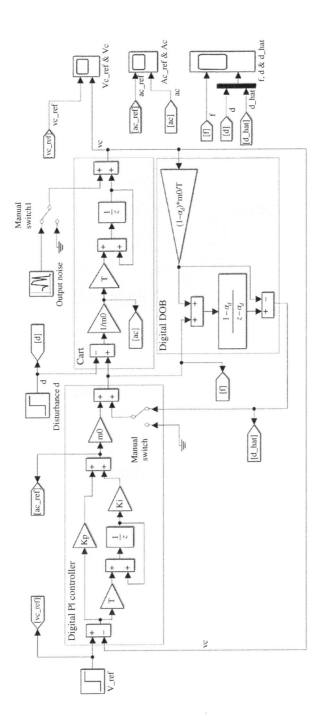

Figure 5.16 Simulink model of digital PI velocity control system.

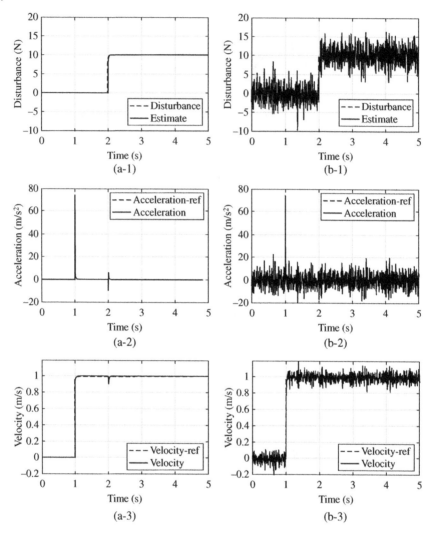

Figure 5.17 Simulation of digital PI velocity control system. (a-1) Disturbance and estimate (without noise), (a-2) acceleration (without noise), (a-3) velocity (without noise), (b-1) disturbance and estimate (with noise), (b-2) acceleration (with noise), and (b-3) velocity (with noise).

low-pass filter section of the continuous system. The waveforms in Column (a) in Figure 5.17 are almost identical to those in Figure 2.11(a). The waveforms in Column (b) in Figure 5.17 are close to those in Figure 2.23(a), but the effect of noise is slightly reduced.

5.4 Identity High-order Digital Disturbance Observer

Consider the disturbance $d(k)$ as a $p - 1$-order system $(p > 1)$ and express it by Equations (5.32) and (5.33):

$$\eta(k+1) = \begin{bmatrix} \eta_0(k+1) \\ \eta_1(k+1) \\ \vdots \\ \eta_{p-1}(k+1) \end{bmatrix} = \Gamma_d \eta(k) = \Gamma_d \begin{bmatrix} \eta_0(k) \\ \eta_1(k) \\ \vdots \\ \eta_{p-1}(k) \end{bmatrix}$$

$$\Gamma_d = e^{\Gamma T}, \quad \Gamma_d \in R^{p \times p} \tag{5.32}$$

$$d(k) = H\eta(k), \quad H = \begin{bmatrix} 1 & 0 & \cdots & 0 \end{bmatrix} \in R^{1 \times p} \tag{5.33}$$

The control plant is expressed as Equation (5.34):

$$x(k+1) = A_d x(k) + B_d u(k) - B_d d(k) = A_d x(k) + B_d u(k) - B_d H\eta(k) \tag{5.34}$$

Let the state variables be $\bar{x}(k) = \begin{bmatrix} x(k) & \eta(k) \end{bmatrix}^T$ and create Equations (5.35) and (5.36):

$$\bar{x}(k+1) = \begin{bmatrix} x(k+1) \\ \eta(k+1) \end{bmatrix} = \bar{A}_d \bar{x}(k) + \bar{B}_d u(t)$$

$$= \begin{bmatrix} A_d & -B_d H \\ 0 & \Gamma_d \end{bmatrix} \begin{bmatrix} x(k) \\ \eta(k) \end{bmatrix} + \begin{bmatrix} B_d \\ 0 \end{bmatrix} u(k) \tag{5.35}$$

$$y(k) = \bar{C}_d \bar{x}(k) = \begin{bmatrix} C_d & 0 \end{bmatrix} \begin{bmatrix} x(k) \\ \eta(k) \end{bmatrix} \tag{5.36}$$

In this case, if $(\bar{C}_d, \bar{A}_d)$ is observable, then a $p - 1$-order digital disturbance observer can be designed, which can be expressed as Equations (5.37)–(5.39):

$$\hat{\bar{x}}(k+1) = \begin{bmatrix} \hat{x}(k+1) \\ \hat{\eta}(k+1) \end{bmatrix} = \bar{A}_d \hat{\bar{x}}(k) + \bar{B}_d u(k) + \hat{H}(y(k) - \hat{y}(k))$$

$$= \begin{bmatrix} A_d & -B_d H \\ 0 & \Gamma_d \end{bmatrix} \begin{bmatrix} \hat{x}(k) \\ \hat{\eta}(k) \end{bmatrix} + \begin{bmatrix} B_d \\ 0 \end{bmatrix} u(k)$$

$$+ \begin{bmatrix} H_{d1} \\ H_{d2} \end{bmatrix} (y(k) - \hat{y}(k)) \tag{5.37}$$

$$\hat{y}(k) = \bar{C}_d \hat{\bar{x}}(k) = \begin{bmatrix} C_d & 0 \end{bmatrix} \begin{bmatrix} \hat{x}(k) \\ \hat{\eta}(k) \end{bmatrix} \tag{5.38}$$

$$\hat{d}(k) = H\hat{\eta}(k) \tag{5.39}$$

Example 5.4.1 (Higher order digital disturbance observer with position observation.) Consider the second-order disturbance observer of the cart model, observer($p = 2$). Given the order $n = 2$, the number of control inputs $m = 1$, and the number of outputs $l = 1$, we have $H = [1, 0]$ and $\Gamma = \begin{bmatrix} 0 & 1 \\ 0 & 0 \end{bmatrix}$. However, to derive Γ_d, we used $\Gamma_d = exp(\Gamma T)$.

An illustrative program and a Simulink model are in List 5.4 and Figure 5.18, respectively. The disturbance estimation waveform shown in Figure 5.19a is the

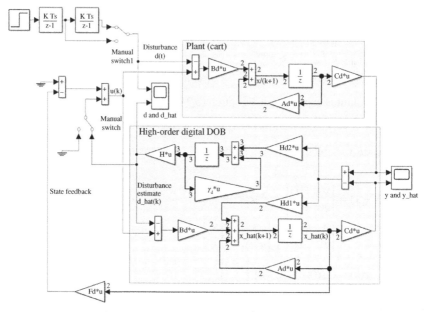

Note: The initial value x_ob0 of DOB should be set to 0 or x0 in the m-file as appropriate.

Figure 5.18 Simulink model of higher order digital DOB.

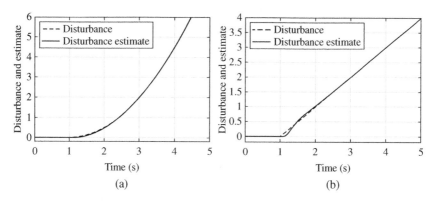

Figure 5.19 Simulation example of higher order digital DOB. (a) Second-order disturbance and (b) first-order disturbance.

estimated waveform of a **second-order disturbance**, and Figure 5.19(b) is the estimated waveform of a **first-order disturbance**.

List 5.4: High-order digital disturbance observer design.

```
1   %% Pysical parameters
2   m=2.0;c=0.10; % Mass 2.0 kg, Viscous friction coefficient 0.10 N/(m/s)
3   A=[0 1; 0 -c/m];B=[0;1/m];C=[1 0];D=0;
4   x0=[0.1;0.0]; % Initial value position xc(0)=0.1m velocity vc(0)=0.0m/s
5   [n,m]=size(B); % Order and number of inputs次数と入力数
6
7   %% Digitization and controller design
8   T=0.01;[Ad,Bd,Cd,Dd]=c2dm(A,B,C,D,T);
9
10  %% Controller design:
11  pole=[0.98,0.99];Fd=place(Ad,Bd,pole);% Gain design with
12  % pole assignment method
13
14  %% Assumption of disturbance and simulation time
15  p=2; % dimension of disturbance to be determined by designer
16  % p=0 to 4 is observable
17  tend=5; % Simulation time
18  dis_step=1.0; % Set the magnitude of the stepwise disturbance
19
20  %% p-th order DOB design and setup for evaluation
21  N=Bd;% Assume matching condition
22  H=1;Gamma=zeros(p+1,p+1);for i=1:p H=[H,0];end;
23  Gamma(1:p,2:p+1)=eye(p);% Create Gamma
24  Gamma_d=expm(Gamma*T) % The expm(Gamma*T) in the second half is the key
25  Abar=[Ad,-N*H;zeros(p+1,n),Gamma_d];
26  Bbar=[Bd;zeros(p+1,1)];Cbar=[Cd,zeros(1,p+1)];
27  %
28  pole_ob=[];base_of_pole=0.85;delta_pole=0.02;
29  for i=0:n+p pole_ob=[pole_ob,base_of_pole+delta_pole*i]; end;
30  Htemp=place(Abar',Cbar',pole_ob);Hob=Htemp';% Design gain by pole assignment method
31  Hd1=Hob(1:n);Hd2=Hob(n+1:n+p+1);% Gain for state and disturbance estimates
32  x_ob0=x0; % x ob0 is the initial value of x hat
33  % Compare the case with zero value and the case with x0
34
35  %% Simulation
36  open_system('sim_Figure_5_18_and_5_19_high_order_digital_dob');
37  set_param('sim_Figure_5_18_and_5_19_high_order_digital_dob','WideLines','on');
38  set_param('sim_Figure_5_18_and_5_19_high_order_digital_dob','ShowLineDimensions','on');
39  z=sim('sim_Figure_5_18_and_5_19_high_order_digital_dob');
```

References

1 Yuhki Kosaka, Akira Shimada: Motion Control for Articulated Robots Based on Accurate Modeling, The 8th IEEE International Workshop on Advanced Motion Control (AMC'04), 535–540, 2004.

2 Akira Shimada, Chaisamorn Yongyai: Motion control of inverted pendulum robots using a Kalman filter based disturbance observer, Transactions of SICE JCMSI, Vol. 2, No. 1, 50–55, 2009.

6

Disturbance Observer of Vibrating Systems

Mechatronics systems are subject to vibration. A vibrating mechatronics system has a single or multiple **resonance frequencies**, and when the vibration is excited, it mainly generates vibration at the resonance frequencies. Vibrating mechatronics systems are often represented using mathematical models with multiple rigid bodies connected by elastic parts, the simplest example being the **two-inertia systems**.

6.1 Modeling of the Two-inertia System

Consider the two-inertia rotating system in Figure 6.1 as the most basic **vibration system**. The input shaft has rotation angle θ_1, rotation velocity ω_1, torque τ, and moment of inertia J_1; the output shaft has rotation angle θ_2, rotation velocity ω_2, and moment of inertia J_2. Furthermore, the two are connected by an elastic part with a coefficient of elasticity k and a coefficient of viscous friction c.

Moreover, assume that a disturbance torque τ_{d1} is applied to the input shaft, or a disturbance torque τ_{d2} is applied to the output shaft to prevent rotation.[1] This model corresponds, for example, to a system in which a motor drives a mechatronics system on the output shaft via an elastic reducer and is common to many rotating mechatronics systems. Figure 6.2a or b is often used for the transfer function block diagram.

The equations of motion are expressed by Equations (6.1) and (6.2):

$$J_1\ddot{\theta}_1 + c(\omega_1 - \omega_2) + k(\theta_1 - \theta_2) = \tau - \tau_{d1} \tag{6.1}$$

$$J_2\ddot{\theta}_2 - c(\omega_1 - \omega_2) - k(\theta_1 - \theta_2) = -\tau_{d2} \tag{6.2}$$

---

1 Note that this is different from the output disturbance.

Disturbance Observer for Advanced Motion Control with MATLAB/Simulink, First Edition. Akira Shimada.
© 2023 The Institute of Electrical and Electronics Engineers, Inc. Published 2023 by John Wiley & Sons, Inc.
Companion website: www.wiley.com/go/disturbanceobserver

Input shaft side   Output shaft side

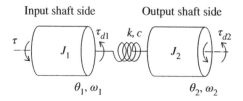

Figure 6.1 Two-inertia system.

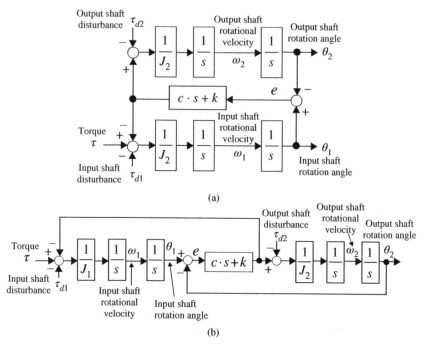

Figure 6.2 Block diagram of the two-inertia system transfer function. (a) Block diagram arranged vertically and (b) block diagram arranged horizontally.

Whether the angle sensor is mounted on the input or output shaft depends on the purpose of use and constraint conditions of the mechatronics system.[2] The torque τ on the input shaft and the rotation angle with respect to the disturbances τ_{d1} and τ_{d2} are given using the transfer function, $\theta_1(s) = G_{11}(\tau - \tau_{d1}) + G_{12}\tau_{d2}$ and $\theta_2(s) = G_{21}(\tau - \tau_{d1}) + G_{22}\tau_{d2}$, and specifically given by Equation (6.3):

$$\begin{bmatrix} \theta_1(s) \\ \theta_2(s) \end{bmatrix} = \begin{bmatrix} \dfrac{N_{11}(s)}{D(s)} & \dfrac{N_{12}(s)}{D(s)} \\ \dfrac{N_{21}(s)}{D(s)} & \dfrac{N_{22}(s)}{D(s)} \end{bmatrix} \begin{bmatrix} \tau - \tau_{d1} \\ \tau_{d2} \end{bmatrix} \tag{6.3}$$

---

2 The former, where input and observation output are at the same point, is called **collocation**, and the latter is called **noncollocation**.

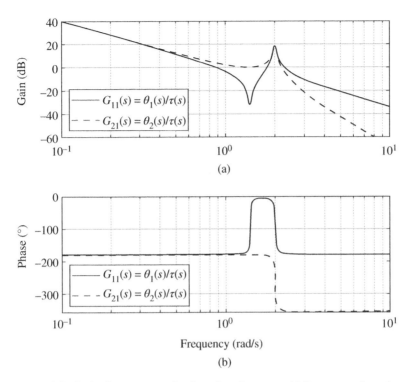

Figure 6.3 Bode diagram example of two inertia system. (a) Represents the gain diagram and (b) represents the phase diagram.

However, if $D(s) = J_1 J_2 s^4 + (J_1 + J_2) c s^3 + (J_1 + J_2) k s^2$,
$N_{11}(s) = J_1 s^2 + cs + k$, $N_{12} = N_{21} = -(cs + k)$, and $N_{22}(s) = J_2 s^2 + cs + k$. Note that if the torque $\tau = \tau_{d1}$, then τ_{d1} can be directly canceled out, whereas τ_{d2} cannot be directly canceled out by τ because its position is far from the input position.

Additionally, the **relative order** between the denominator and numerator of $G_{11}(s) = N_{11}(s)/D(s) = \theta_1(s)/\tau(s)$ is two (=fourth order-second order), the relative order between the denominator and numerator of $G_{21}(s) = N_{21}(s)/D(s) = \theta_2(s)/\tau(s)$ is three (=fourth order-first order).[3] If the phase delay is considerable, the control system closes to be unstable and difficult to control.

The Bode diagram example of $G_{11}(s)$ and $G_{21}(s)$ is shown in Figure 6.3. The G_{11} waveform shows the antiresonance and resonance characteristics, and the phase goes down to $-180°$ once, while the phase goes up to $-180°$. In contrast, the waveform of G_{21} shows only the resonance property, and the phase does not advance to

---

3 If the coefficient c-value is almost zero, the order difference number is fourth order (=fourth - zeroth order). Then, the phase converges to $(-90° \times$ order difference number) in the high-frequency range.

$-360°$. The G_{21} waveform shows only the resonance Characteristics, and no phase advances to $-360°$.

6.2 Vibration Suppression Control in Transfer Function Representation

Even if the control plant is a two-inertia system, we dare to assume that it is a one-inertia system with approximation. For example, suppose what happens if we create a velocity PI control system using a disturbance observer with the same configuration as Figure 2.3b.[4]

Example 6.2.1 (Numerical example with simple implementation.)   An example program is shown in List 6.1; Figure 6.4 shows the Simulink model; and the waveform is shown in Figure 6.5.

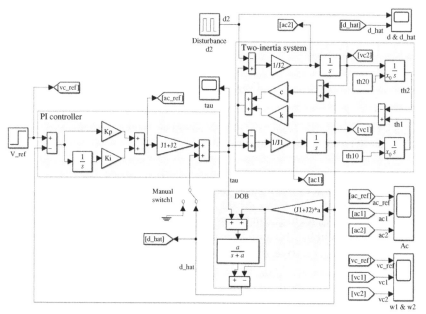

Figure 6.4 Simulink model of velocity PI control system using simple disturbance observer (DOB) for a two-inertia system.

---

4 In controlling a mechatronics system where the decelerator has nonnegligible elasticity, this is often the case without the designer realizing it. For example, assume that the moment of inertia is the sum of the input and output shafts, $J_1 + J_2$. Furthermore, the disturbance is assumed to be on the output shaft only.

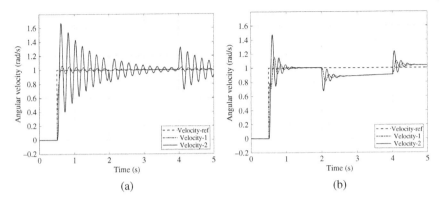

Figure 6.5 Simulation results of velocity PI control using simple DOB for a two-inertia system. (a) Velocity (SW = on) and (b) velocity (SW = off).

List 6.1: DOB and disturbance suppression for two inertia systems.

```
1   %% Physical parameters
2   % Rotation angle:th1,th2, rotation velocity:w1,w2
3   J1=1.0;J2=2−J1;k=1e3;c=0.1; % Moment of inertia
4   % elasticity coefficient,viscosity coefficient
5   tau_d2=10;T_cyc=2; % magnitude and period of disturbance
6   % to output shaft
7
8   %% DOB and PI controller design
9   T=0.01;a=1/T; % Time constant of DOB
10  zeta=7.5;wn=2.5; % Damping constant and response frequency
11  % of control system
12  Kp=2*zeta*wn;Ki=wn^2; % PI controller gains
13
14  %% Simulation
15  th10=0;th20=th10; % Initial value
16  tend=5; % Simulation time
17  open_system('sim_Figure_6_4_and_6_5_two_mass_control_with_simple_DOB');
18  y=sim('sim_Figure_6_4_and_6_5_two_mass_control_with_simple_DOB');
```

The left column in Figure 6.5 shows an example of turning on the SW in Figure 6.4 and returning the disturbance estimate $\hat{d}$, while the right column shows an example of turning it off. The input shaft velocity follows the reference value, but the output shaft velocity in the left column (SW on) is oscillatory. The velocity waveform in the right column (SW off) is less oscillatory, but a steady deviation remains when the disturbance is nonzero.

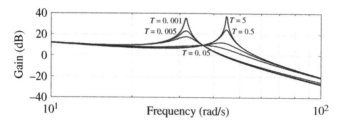

Figure 6.6 Output shaft gain diagram of simple speed control system with DOB.

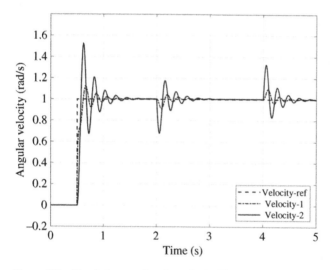

Figure 6.7 Simulation results of a velocity PI control system ($T = 0.05$ seconds) using a simple DOB for a two-inertia system.

Various methods have been proposed to improve the characteristics of such phenomena, such as resonance ratio control [1, 2]. For example, readjusting the time constant, T, of the disturbance observer can produce a vibration suppression effect [3]. The open-loop gain diagram of the control system for various T values is shown in Figure 6.6. SW is on. The resonant characteristics vary significantly with the T value. The resonant peak almost disappears between small and large values. Figure 6.7 shows the simulation results obtained using the T values (=0.05 seconds) in the valleys. Although insufficient, there is an improvement in the characteristics, and the controller suppresses the steady-state error.

6.3 Disturbance Observer and Stabilization for Two-inertia Systems

Consider an observer that estimates the input shaft disturbance τ_{d1}, the output shaft disturbance τ_{d2}, and a vibration suppression controller design simultaneously. Here, assume that a mounted sensor can observe only the input shaft's rotation angle.

6.3.1 Observer to Estimate Input Shaft Disturbance τ_{d1}

The state and output equations are created from Equations (6.1) and (6.2) using the state variables $x = [\theta_1, \theta_2, \omega_1, \omega_2]^T$, input $u = \tau$, output $y = \theta_1$, and assuming that there is input shaft disturbance $d = \tau_{d1}$ and no output shaft disturbance. Equations (6.4) and (6.5) are obtained in the form $\dot{x} = Ax + B_1 u - B_1 d$ and $y = C_1 x$:

$$
\begin{bmatrix} \dot{\theta}_1 \\ \dot{\theta}_2 \\ \dot{\omega}_1 \\ \dot{\omega}_2 \end{bmatrix} = \begin{bmatrix} 0 & 0 & 1 & 0 \\ 0 & 0 & 0 & 1 \\ -\frac{k}{J_1} & \frac{k}{J_1} & -\frac{c}{J_1} & \frac{c}{J_1} \\ \frac{k}{J_2} & -\frac{k}{J_2} & \frac{c}{J_2} & -\frac{c}{J_2} \end{bmatrix} \begin{bmatrix} \theta_1 \\ \theta_2 \\ \omega_1 \\ \omega_2 \end{bmatrix} + \begin{bmatrix} 0 \\ 0 \\ \frac{1}{J_1} \\ 0 \end{bmatrix} \tau
$$

$$
- \begin{bmatrix} 0 & 0 & \frac{1}{J_1} & 0 \end{bmatrix}^T \tau_{d1} \tag{6.4}
$$

$$
y = \begin{bmatrix} 1 & 0 & 0 & 0 \end{bmatrix} \begin{bmatrix} \theta_1 & \theta_2 & \omega_1 & \omega_2 \end{bmatrix}^T \tag{6.5}
$$

We can obtain Equations (6.10) and (6.7), if we assume that $\dot{d} = \dot{\tau}_{d1} = 0$ and design the observer after creating the extended system:

$$
\begin{bmatrix} \dot{\hat{\theta}}_1 \\ \dot{\hat{\theta}}_2 \\ \dot{\hat{\omega}}_1 \\ \dot{\hat{\omega}}_2 \\ \dot{\hat{\tau}}_{d1} \end{bmatrix} = \begin{bmatrix} 0 & 0 & 1 & 0 & 0 \\ 0 & 0 & 0 & 1 & 0 \\ -\frac{k}{J_1} & \frac{k}{J_1} & -\frac{c}{J_1} & \frac{c}{J_1} & -\frac{1}{J_1} \\ \frac{k}{J_2} & -\frac{k}{J_2} & \frac{c}{J_2} & -\frac{c}{J_2} & 0 \\ 0 & 0 & 0 & 0 & 0 \end{bmatrix} \begin{bmatrix} \hat{\theta}_1 \\ \hat{\theta}_2 \\ \hat{\omega}_1 \\ \hat{\omega}_2 \\ \hat{\tau}_{d1} \end{bmatrix} + \begin{bmatrix} 0 \\ 0 \\ \frac{1}{J_1} \\ 0 \\ 0 \end{bmatrix} \tau
$$

$$
+ \begin{bmatrix} H_{11} & H_{12} & H_{13} & H_{14} & H_2 \end{bmatrix}^T (y - \hat{y}) \tag{6.6}
$$

$$
\hat{y} = \begin{bmatrix} 1 & 0 & 0 & 0 & 0 \end{bmatrix} \begin{bmatrix} \hat{\theta}_1 & \hat{\theta}_2 & \hat{\omega}_1 & \hat{\omega}_2 & \hat{\tau}_{d1} \end{bmatrix}^T \tag{6.7}
$$

Example 6.3.1 (Numerical example of an observer estimating the input shaft disturbance τ_{d1}.) An example program is shown in List 6.2; Figure 6.8 shows the Simulink model; and the estimated waveform is shown in Figure 6.9. However,

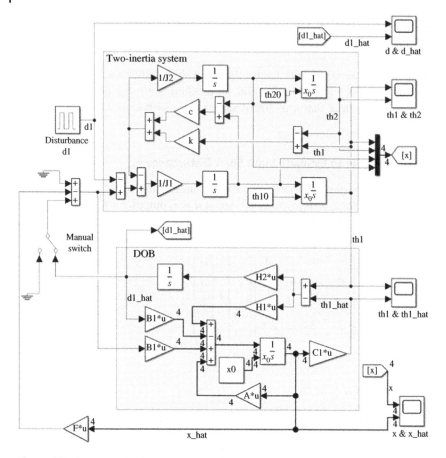

Figure 6.8 Example of a Simulink model of a control system for estimating input shaft disturbances.

in this example, the observer gain was designed using the design method of the optimal regulator,[5] since no suitable pole assignment could be found. Meanwhile, when the modulus of elasticity k was stiffened to 1000 N m/rad, a stable observer could be designed using the pole assignment method.[6] We tried to converge to zero for the control system using LQ[7] control.

---

5 Sometimes referred to as the optimal observer because it is equivalent to the stationary Kalman filter.

6 Even if you can construct a seemingly clean theory, you may encounter such situations in various examples. In order to study the effect of different k values, we would like to compare the observable Gramian values, but the observable Gramian can only be calculated if all the poles of A_{bar} are negative system.

7 See Appendix A.3.7 about LQ control.

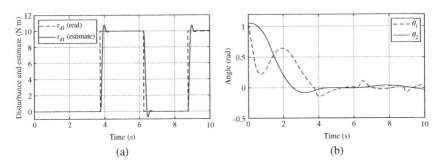

Figure 6.9 Simulation results for a two-inertia system with input shaft disturbance.
(a) Output shaft disturbance and estimate and (b) input and output shaft angle.

List 6.2: DOB and disturbance suppression for two inertia systems.

```
1   %% Physical parameters
2   % Rotation angle:th1,th2, rotation velocity:w1,w2
3   J1=1.0;J2=2−J1;k=1;c=0.01; % Moment of inertia
4   % elasticity coefficient,viscosity coefficient
5   tau_d1=10;T_cyc=5; % magnitude and period of disturbance
6
7   %% Create state space model
8   A=[0, 0, 1,0;
9      0, 0, 0,1;
10     −k/J1, k/J1,−c/J1, c/J1;
11      k/J2,−k/J2, c/J2,−c/J2];
12  B1=[0;0;1/J1;0];
13  C1=[1,0,0,0]; % C1 corresponds to input−shaft observation
14
15  %% Create extended system and design DOB
16  A_bar=[A,−B1;zeros(1,5)];C1_bar=[C1,0];
17  Q_ob=diag([1e2,1e2,1e1,1e1,1e8]);R_ob=1;
18  H_temp=lqr(A_bar',C1_bar',Q_ob,R_ob); % DOB design with LQ method
19  H1=H_temp(1:4)';H2=H_temp(5); % Components 1 to 4 are x−estimation,
20  % 5th component is disturbance estimation gain
21
22  %% Control system design
23  Q=diag([1e1,1e3,1e2,1e2]);R=1;F=lqr(A,B1,Q,R);
24
25  %% Simulation
26  th10=pi/3;th20=th10;x0=[th10;th20;0;0]; % Initial values
27  tend=10;
28  open_system('sim_Figure_6_8_and_6_9_DOB_for_two_inertia_system_with_taud1');
```

```
29   set_param('sim_Figure_6_8_and_6_9_DOB_for_two_inertia_system_with_taud1','WideLines','on');
30   set_param('sim_Figure_6_8_and_6_9_DOB_for_two_inertia_system_with_taud1',
         'ShowLineDimensions','on');
31   y=sim('sim_Figure_6_8_and_6_9_DOB_for_two_inertia_system_with_taud1');
```

The disturbance estimate $\hat{\tau}_{d1}$ converges to the input shaft disturbance τ_{d1}. However, there is still a small overshoot. However, the angular waveform can quickly converge to zero though slightly disturbed by the disturbance. If the positive feedback of the disturbance is removed by switching the Manual Switch in the left part of the diagram, the disturbance estimation function remains unchanged, but the rotation angle does not converge to zero, and an offset remains.

6.3.2 Observer to Estimate Output Shaft Disturbance τ_{d2}

Assume state variables $x = [\theta_1, \theta_2, \omega_1, \omega_2]^T$, input $u = \tau$, output $y = \theta_1$, output shaft disturbance $d = \tau_{d2}$, and no input shaft disturbance. To express the corresponding plant, Equations (6.8) and (6.9) utilize the form of $\dot{x} = Ax + B_1u - B_2d$, $y = C_1x$:[8]

$$\begin{bmatrix} \dot{\theta}_1 \\ \dot{\theta}_2 \\ \dot{\omega}_1 \\ \dot{\omega}_2 \end{bmatrix} = \begin{bmatrix} 0 & 0 & 1 & 0 \\ 0 & 0 & 0 & 1 \\ -\frac{k}{J_1} & \frac{k}{J_1} & -\frac{c}{J_1} & \frac{c}{J_1} \\ \frac{k}{J_2} & -\frac{k}{J_2} & \frac{c}{J_2} & -\frac{c}{J_2} \end{bmatrix} \begin{bmatrix} \theta_1 \\ \theta_2 \\ \omega_1 \\ \omega_2 \end{bmatrix} + \begin{bmatrix} 0 \\ 0 \\ \frac{1}{J_1} \\ 0 \end{bmatrix} \tau$$
$$- \begin{bmatrix} 0 & 0 & 0 & \frac{1}{J_2} \end{bmatrix}^T \tau_{d2} \tag{6.8}$$

$$y = \begin{bmatrix} 1 & 0 & 0 & 0 \end{bmatrix} \begin{bmatrix} \theta_1 & \theta_2 & \omega_1 & \omega_2 \end{bmatrix}^T \tag{6.9}$$

After all, if we assume that $\dot{d} = \dot{\tau}_{d2} = 0$ and design the observer after creating the extended system, we can get Equations (6.10) and (6.11):

$$\begin{bmatrix} \dot{\hat{\theta}}_1 \\ \dot{\hat{\theta}}_2 \\ \dot{\hat{\omega}}_1 \\ \dot{\hat{\omega}}_2 \\ \dot{\hat{\tau}}_{d2} \end{bmatrix} = \begin{bmatrix} 0 & 0 & 1 & 0 & 0 \\ 0 & 0 & 0 & 1 & 0 \\ -\frac{k}{J_1} & \frac{k}{J_1} & -\frac{c}{J_1} & \frac{c}{J_1} & 0 \\ \frac{k}{J_2} & -\frac{k}{J_2} & \frac{c}{J_2} & -\frac{c}{J_2} & -\frac{1}{J_2} \\ 0 & 0 & 0 & 0 & 0 \end{bmatrix} \begin{bmatrix} \hat{\theta}_1 \\ \hat{\theta}_2 \\ \hat{\omega}_1 \\ \hat{\omega}_2 \\ \hat{\tau}_{d2} \end{bmatrix} + \begin{bmatrix} 0 \\ 0 \\ \frac{1}{J_1} \\ 0 \\ 0 \end{bmatrix} \tau$$
$$+ \begin{bmatrix} H_{11} & H_{12} & H_{13} & H_{14} & H_2 \end{bmatrix} (y - \hat{y}) \tag{6.10}$$

---

8 The vertical vectors span several lines, so we have used transposed expressions for the horizontal vectors where appropriate.

$$\hat{y} = \begin{bmatrix} 1 & 0 & 0 & 0 \vdots 0 \end{bmatrix} \begin{bmatrix} \hat{\theta}_1 & \hat{\theta}_2 & \hat{\omega}_1 & \hat{\omega}_2 \vdots \hat{\tau}_{d2} \end{bmatrix}^T \qquad (6.11)$$

Example 6.3.2 (Numerical example of an observer estimating the output shaft disturbance τ_{d2}.) An example program is shown in List 6.3, and the Simulink model and estimated waveforms are shown in Figure 6.11. Contrary to the previous example, the observer was designed using the pole assignment method.

List 6.3: DOB and disturbance suppression for two-inertia system with d2 disturbance.

```
1    %% Physical parameters
2    % Rotation angle:th1,th2, rotation velocity:w1,w2
3    J1=1.0;J2=2-J1;k=1;c=0.01; % Moment of inertia
4    % elasticity coefficient,viscosity coefficient
5    tau_d2=10;T_cyc=5; % magnitude and period of disturbance
6
7    %% Create state space model
8    A=[0, 0, 1,0;
9       0, 0, 0,1;
10      -k/J1, k/J1,-c/J1, c/J1;
11      k/J2,-k/J2, c/J2,-c/J2];
12   B1=[0;0;1/J1;0];B2=[0;0;0;1/J2]; % B2 corresponds to output-shaft disturbance
13   C1=[1,0,0,0]; % C1 corresponds to input-shaft observation
14
15   %% Create extended system and design DOB
16   A_bar=[A,-B2;zeros(1,5)];C1_bar=[C1,0];
17   pole_dob=[-30,-32,-34,-36,-39];
18   H_temp=place(A_bar',C1_bar',pole_dob); % DOB design with pole assignment
19   H1=H_temp(1:4)';H2=H_temp(5); % Components 1 to 4 are x-estimation,
20   % 5th component is disturbance estimation gain
21
22   %% Control system design
23   Q=diag([1e1,1e3,1e2,1e2]);R=1;
24   F=lqr(A,B1,Q,R);
25
26   %% Simulation
27   th10=pi/3;th20=th10;x0=[th10;th20;0;0];% Initial values
28   tend=20;
29   open_system('sim_Figure_6_10_and_6_11_DOB_for_two_inertia_system_with_taud2');
30   set_param('sim_Figure_6_10_and_6_11_DOB_for_two_inertia_system_with_taud2','WideLines','on');
31   set_param('sim_Figure_6_10_and_6_11_DOB_for_two_inertia_system_with_taud2',
             'ShowLineDimensions','on');
32   y=sim('sim_Figure_6_10_and_6_11_DOB_for_two_inertia_system_with_taud2');
```

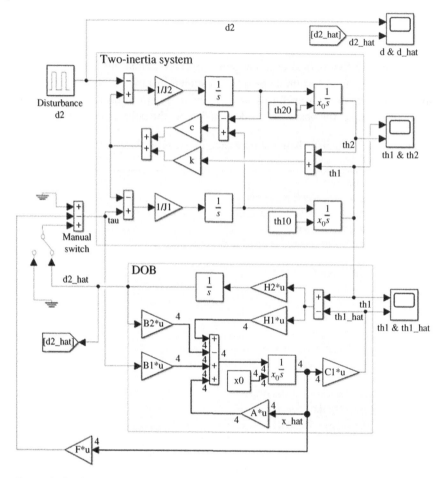

Figure 6.10 Simulink model of control system to estimate output shaft disturbance.

A pulsed disturbance to the output shaft, shown in the upper left corner of Figure 6.10, inserts into the output shaft. However, in the observer, one can see that the input goes through B_1, and the disturbance estimate goes through B_2 to the state estimation part. Figure 6.11a shows the output shaft disturbance and the estimated waveform that follows the disturbance without overshoot. This is a good example of how disturbance estimation is possible even if $B_2 \neq B_1$, as long as it is observable.[9] Figure 6.11b shows the rotation angle waveforms of the input and output shafts. The estimated value cannot be suppressed by positive feedback since the disturbance is not the input shaft disturbance. Although the system was

---

9 When $B_2 = \alpha B_1$ ($\alpha \neq 0 \in R$) is true, we say that "the matching condition is satisfied."

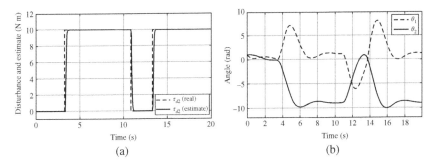

Figure 6.11 Simulation results with output shaft disturbance in a two-inertia system. (a) Output shaft disturbance and estimate and (b) input and output shaft angle waveforms.

successfully stabilized, it did not converge to zero. An example of disturbance suppression control for such a case is discussed in the following section 6.4.

6.4 Servo System with DOB for Two-inertia Systems

This section deals with an angle servo system. We can assume some variety where disturbance enters the input or output shaft, and the controller can observe the output θ_1 or θ_2. Since there are various alternatives, let us summarize them in Table 6.1.

Next, depending on whether one wants to control the input or output shaft angle, the variable to be compared with the reference value should be replaced. In practice, there are four options: the actually observed output θ_1 or θ_2, or the estimated value $\hat{\theta}_1$ or $\hat{\theta}_2$. Overall, there are $4 \times 6 = 24$ possible options. However, this section assumes that the controller uses only the control input and the observed value θ_1 for the disturbance observer design, the disturbance is either input or output shaft, and can use two outputs $\hat{\theta}_1$ and $\hat{\theta}_2$ for the servo system design.

Table 6.1 Kinds of outputs and disturbances

Observed output	Input shaft disturbance	Output shaft disturbance
θ_1	Exist	Not exist
	Not exist	Exist
	Exist	Exist
θ_2	Exist	Not exist
	Not exist	Exist
	Exist	Exist

6.4.1 Input Shaft Servo System Considering Input Shaft Disturbance τ_{d1}

There is a case where the input shaft of a two-inertia system is the main mechatronics part, and the output shaft is an accessory part. Here, we assume that the angle sensor located on the input shaft's angle is to be controlled. Meanwhile, there is another case where the output shaft's angle control is the main purpose of using the mechatronics system. For example, the reduction gear between the motor and the output is elastic, and the sensor is mounted on the motor shaft. Here, even if the angle control of the input shaft is successful, nonnegligible vibration may be generated on the output shaft. Then, using a control gain that speeds up the convergence of θ_1, θ_2 generates significant vibration in the simulation. In such cases, using the reduced control gain or reduced observer gain may be one choice to implement a stable but slow response characteristic.

Example 6.4.1 (Numerical example of input shaft servo system.) The observation matrix $C_1 = [1, 0, 0, 0]$ outputting θ_1 and the observation matrix $C_2 = [0.1, 0, 0, 0]$ outputting θ_2 are set to List 6.4, and the example program is set to the Simulink. The model is shown in Figure 6.12, and the response waveforms are shown in Figure 6.13.

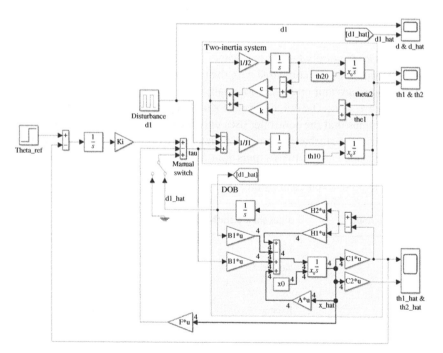

Figure 6.12 Simulink model of input shaft servo system considering input shaft disturbance.

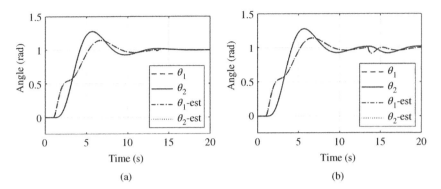

Figure 6.13 Simulation of input shaft servo system considering input shaft disturbance. (a) Angle (SW = on) and (b) angle (SW = off).

List 6.4:   Servo system design for two inertia system with input shaft disturbance.

```
1   %% Physical parameters
2   % Rotation angle:th1,th2, rotation velocity:w1,w2
3   J1=1.0;J2=2-J1;k=1;c=0.001; % Moment of inertia
4   % elasticity coefficient,viscosity coefficient
5   tau_d1=50;T_cyc=18; % magnitude and period of disturbance
6
7   %% Create state space model
8   A=[0, 0, 1,0;
9      0, 0, 0,1;
10     -k/J1, k/J1,-c/J1, c/J1;
11      k/J2,-k/J2, c/J2,-c/J2];
12  B1=[0;0;1/J1;0];B2=[0;0;0;1/J2]; % B2 corresponds to output-shaft disturbance
13  C1=[1,0,0,0]; % C1 corresponds to input-shaft observation
14  C2=[0,1,0,0]; % C2 corresponds to output-shaft observation
15
16  %% Create extended system and design DOB
17  A_bar=[A,-B1;zeros(1,5)];C1_bar=[C1,0];
18  Q_ob=diag([1e2,1e2,1e1,1e1,1e12]);R_ob=1;
19  H_temp=lqr(A_bar',C1_bar',Q_ob,R_ob); % DOB design with pole assignment
20  H1=H_temp(1:4)';H2=H_temp(5); % Components 1 to 4 are x-estimation,
21  % 5th component is disturbance estimation gain
22
23  %% Servo Control system design
24  A_tilde=[A,zeros(4,1);-C1,0];B_tilde=[B1;0];
25  Q=diag([1e2,1e2,1e4,1e5,1e5]);R=1;
26  F_tilde=lqr(A_tilde,B_tilde,Q,R);F=F_tilde(1:4);Ki=-F_tilde(5);
27
28  %% Simulation
29  th10=0;th20=th10;x0=[th10;th20;0;0]; % Initial values
30  tend=20;
31  open_system ('sim_Figure_6_12_and_6_13_two_inertia_servo_taud1');
```

```
32   set_param ('sim_Figure_6_12_and_6_13_two_inertia_servo_taud1','WideLines','on');
33   set_param ('sim_Figure_6_12_and_6_13_two_inertia_servo_taud1','ShowLineDimensions','on');
34   y=sim ('sim_Figure_6_12_and_6_13_two_inertia_servo_taud1');
```

Figures 6.13 (a) and (b) show examples with and without positive feedback, respectively. In Figure 6.13 (a), disturbance suppression is observed and all four waveforms converge to the reference value compared to Figure 6.13 (b).

6.4.2 Output Shaft Servo System Considering Output Shaft Disturbance τ_{d2}

Assume a mechatronics system where the output shaft of a two-inertia system is the primary mechatronics part, and an angle sensor can only be mounted on the input shaft. Then, suppose that we want to control the output shaft angle even though the angle sensor can only be mounted on the input shaft. In the servo system design process, the C_2 matrix is used assuming that θ_2 can be observed, although it cannot actually observe it. In the Simulink model introduced here, not θ_2 but $\hat{\theta}_2$ is connected to the feedback part for the servo control.[10]

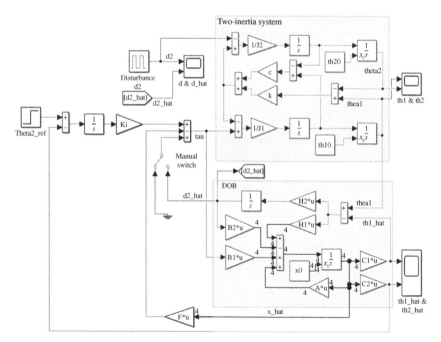

Figure 6.14 Simulink model of output shaft servo system considering output shaft disturbance.

---

10 In a general servo system, the design does not use output estimate but the actual observed output. However, the output shaft servo system can be realized without using an output shaft sensor in this example.

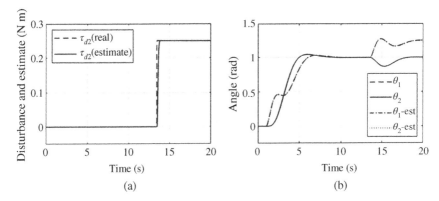

Figure 6.15 Simulation of output shaft servo system. (a) Output shaft disturbance and estimated and (b) angle waveform.

Example 6.4.2 (Numerical example of output shaft servo system.) An example program is shown in List 6.5, and the Simulink model and response waveforms are shown in Figures 6.14 and 6.15, respectively.

List 6.5: Servo system design for two inertia system with output shaft disturbance.

```
1   %% Physical parameters
2   % Rotation angle:th1,th2, rotation velocity:w1,w2
3   J1=1.0;J2=2−J1;k=1;c=0.0001;% Moment of inertia
4   % elasticity coefficient,viscosity coefficient
5   tau_d2=0.25;T_cyc=18; % magnitude and period of disturbance
6
7   %% Create state space model
8   A=[0, 0, 1,0;
9      0, 0, 0,1;
10     −k/J1, k/J1,−c/J1, c/J1;
11     k/J2,−k/J2, c/J2,−c/J2];
12  B1=[0;0;1/J1;0];B2=[0;0;0;1/J2]; % B2 corresponds to output−shaft disturbance
13  C1=[1,0,0,0]; % C1 corresponds to input−shaft observation
14  C2=[0,1,0,0]; % C2 corresponds to output−shaft observation
15
16  %% Create extended system and design DOB
17  A_bar=[A,−B2;zeros(1,5)];C1_bar=[C1,0];
18  pole_dob=[−30,−32,−34,−36,−39];
19  H_temp=place(A_bar',C1_bar',pole_dob); % 極配置でのDOB設計
20  H1=H_temp(1:4)';H2=H_temp(5); % Components 1 to 4 are x−estimation,
21  % 5th component is disturbance estimation gain
22
23  %% Servo Control system design
24  A_tilde=[A,zeros(4,1);−C2,0];B_tilde=[B1;0];
25  Q=diag([1e2,1e2,1e4,1e5,1e5]);R=1;
```

```
26   F_tilde=lqr(A_tilde,B_tilde,Q,R);F=F_tilde(1:4);Ki=-F_tilde(5);
27
28   %% Simulation
29   th10=0;th20=th10;x0=[th10;th20;0;0]; % Initial values
30   tend=20;
31   open_system('sim_Figure_6_14_and_6_15_two_inertia_servo_taud2');
32   set_param('sim_Figure_6_14_and_6_15_two_inertia_servo_taud2','WideLines','on');
33   set_param('sim_Figure_6_14_and_6_15_two_inertia_servo_taud2','ShowLineDimensions','on');
34   y=sim('sim_Figure_6_14_and_6_15_two_inertia_servo_taud2');
```

Figure 6.15a shows the same disturbance waveform as in the previous section, Figure 6.15b shows no difference with or without positive feedback, and θ_1 shows a deviation from the reference value of 1 rad without disturbance suppression. The θ_2 and $\hat{\theta}_2$ converge to the reference value of 1 rad. In this example, we also used a value (0.25 N m) that has a smaller order of magnitude than the previous example (60 N m) because it could not resist large disturbances.

References

1 Kenji Kaneko, Shinichi Kondo, Kouhei Ohishi: A Motion Control of Flexible Joint Based on Velocity Estimation, Proceedings of 16th Annual Conference of IEEE Industrial Electronics Society (IECON'90), 279–284, 1990.

2 Kazuaki Yuki, Toshiyuki Murakami, Kouhei Ohnishi: Vibration Control of 2 Mass Resonant System by Resonance Ratio Control, Proceedings of 19th Annual Conference of IEEE Industrial Electronics(IECON '93), 2009–2014, 1993.

3 Akira Shimada: Servo system design considering low-stiffness of robot, Journal of Robotics and Mechatronics, Vol. 8, No. 3, 252–258, 1996.

7

Communication Disturbance Observer

Control systems over communication networks, including the Internet, experience **communication delays**. It is called **"time delay"** in control engineering and is known to impair the control system's stability. The Smith method is a well-known control method [1, 2]. However, problems such as weak robustness and inability to cope with fluctuations in waste time have been reported. Although several solutions have been introduced, including a control method that considers robust stability [2], this book presents an overview of the Smith method, we explain time delay control, citing mainly the "communication disturbance observer (CDOB) " by Natori et al. [3].

7.1 Smith Method Overview

Let us introduce a control system using the **Smith method** for a system with time delay. Figure 7.1 is the basic block diagram of the control system using the Smith method. The disturbance $d(s)$ is depicted in the diagram, but first, let us assume $d(s) = 0$ and consider the effect of the disturbance later. In $P(s) = G(s)e^{-Ls}$, $G(s)$ is the control plant, L is the time delay, and e^{-Ls} is the time delay element.[1] $C(s)$ is called the controller, and $G(s)(1 - e^{-Ls})$ is called the Smith predictor or compensator.

The closed-loop transfer function of a one-input, one-output system in the absence of a Smith predictor is represented by Equation (7.1):

$$G_{c1}(s) = \frac{P(s)C(s)}{1 + P(s)C(s)} = \frac{G(s)e^{-Ls}(s)C(s)}{1 + G(s)e^{-Ls}C(s)} \tag{7.1}$$

---

1 If $G(s)$ has a time delay L_1 on the input and L_2 on the output, then $e^{-L_2 s}G(s)e^{-L_1 s}$, but since the order of the transfer functions is interchangeable, at least in one-input, one-output systems, this book treats it as $L = L_1 + L_2$ and $e^{-L_2 s}G(s)e^{-L_1 s} = G(s)e^{-Ls}$.

Disturbance Observer for Advanced Motion Control with MATLAB/Simulink, First Edition. Akira Shimada.
© 2023 The Institute of Electrical and Electronics Engineers, Inc. Published 2023 by John Wiley & Sons, Inc.
Companion website: www.wiley.com/go/disturbanceobserver

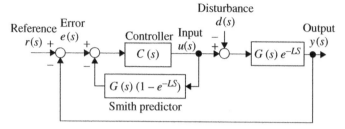

Figure 7.1 Basic block diagram of the Smith method.

It is well known that the system can be destabilized due to a time delay element e^{-Ls} in the denominator. In contrast, with a Smith predictor, the transfer function becomes Equation (7.2):

$$G_{c2}(s) = \frac{C(s)G(s)e^{-Ls}}{1 + C(s)G(s)} = G_{c1}(s)e^{-Ls} \tag{7.2}$$

It is known that using the Smith method eliminates the e^{-Ls} in the denominator of the transfer function, thus eliminating the instability factor, and that the characteristic of the closed-loop characteristic without the time delay is that the time delay is just added. However, for this control method to remain effective, it is necessary that the L of the idle time be known and time invariant. In contrast, the use of a CDOB is expected to handle fluctuations in L.

7.2 Communication Disturbance Observer

The state and output equations of the control plant, including the delay time, are given by Equations (7.3) and (7.4):

$$\dot{x}(t) = Ax(t) + Bu(t - L) \tag{7.3}$$
$$y(t) = Cx(t) \tag{7.4}$$

Equation (7.3) is rewritten somewhat aggressively as Equation (7.5):

$$\dot{x}(t) = Ax(t) + Bu(t) - Bd(t) \tag{7.5}$$

That is, $Bu(t - L) = Bu(t) - Bd(t)$, so Equation (7.6) holds:

$$d(t) = u(t) - u(t - L) \tag{7.6}$$

We can obtain $d(s) = u(s) - u(s)e^{-Ls} = (1 - e^{-Ls})u(s)$ via the Laplace transformation.

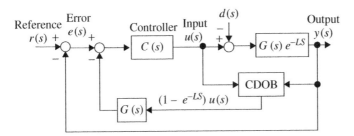

Figure 7.2 Block diagram of the Smith method incorporating a communication disturbance observer.

For the observer design, assuming $\dot{d}(t) = 0$ to simplify the design, a general disturbance observer (DOB) is designed and applied to the control system with the Smith method, which becomes Figure 7.2.[2]

Example 7.2.1 (Comparison of Smith method and CDOB.) We compare the two control systems. The first system is a PID position control system with a Smith predictor. The second is the same system with a CDOB.[3] However, for the actual communication time delay $L = 0.2$ seconds, the estimated delay time, L_{est}, is assumed to be much smaller by a factor of $1/4$. For illustration, we tried a nasty setup in which a step-like disturbance is added at about $2/3$ (14 seconds) of the simulation time.

The program list is shown in List 7.1; the Simulink model of the PID position control system with the Smith predictor is shown in Figure 7.3; and an example

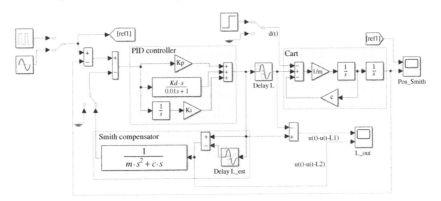

Figure 7.3 Simulink model of control system with Smith method.

---

2 The block labeled CDOB in Figure 7.2 is actually a general DOB for the control plant $G(s)$, which can be designed for both continuous and digital systems at the designer's discretion.
3 A detailed comparative verification is reported in [3]. Appendix A.2.3 introduces the basic information related to PID control.

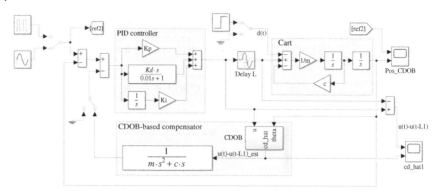

Figure 7.4 Simulink model of control system incorporating CDOB.

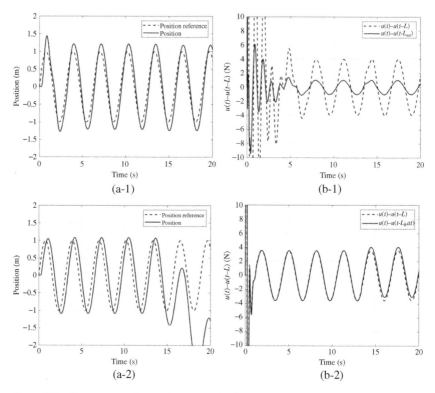

Figure 7.5 Comparison of control system with Smith method and control system with CDOB. (a-1) Position waveform (Smith method), (a-2) position waveform (control with CDOB), (b-1) delay waveform (Smith method), and (b-2) delay (control with CDOB).

utilizing CDOB is shown in Figure 7.4. The position waveforms of the former control system with the Smith predictor are shown in Figure 7.5a-1, and the input delay $u(t) - u(t - L)$ is shown in (b-1). Similarly, the waveforms of the latter control with the CDOB are shown in Figure 7.5a-1,b-1.

List 7.1: Comparison of Smith method and CDOB-based control system.

```
1   %% Physical parameters and PID controller gains
2   m=2; % <ass [kg]
3   L=0.2;L_est=L*0.25;% Real deley:L and delay estimate:L_est
4   w=2;c=1; % Angular frequency of position reference[rad/s]
5   Kp=30;Ki=5;Kd=15; % PID gains
6
7   %% CDOB design in state space representation
8   A=[0 1;0 -c/m];B=[0; 1/m];C=[1,0];
9   Abar=[A,-B;[0,0],0];Cbar=[C,0];
10  pole=[-10,-12, -13]*10;
11  Htemp=place(Abar',Cbar',pole);H=Htemp';
12  H1=H(1:2);H2=H(3);
13
14  %% Simulation
15  tend=20; % Simulation time
16  t_dis=tend*2/3; % Disturbance generation time
17  dis=0.5; % disturbance magnitude
18  open_system('sim_Figure_7_3_7_4_and_7_5_Smith_method_and_CDOB');
19  z=sim('sim_Figure_7_3_7_4_and_7_5_Smith_method_and_CDOB');
```

The ideal time delay control in the Smith predictor combined control can be realized when $L_{est} = L$ and $d = 0$. Then, the position control deviation shown in Figure 7.5a-1 is not small but not as too large as expected, even though the estimated value L_{est} is intentionally shifted significantly from the L value. However, the position control deviation gradually shifts after $t = 14$ seconds, where the disturbance is added. Moreover, in Figure 7.5b-1, the actual delay time deviates significantly from the estimated value.

Meanwhile, in the control with the CDOB, as seen in the Figure 7.5a-2, good delay time control is generally achieved up to $t = 14$ seconds. However, a significant position deviation occurs after a disturbance is introduced. In the waveform of the delay time estimation in the $t = 14$ seconds(Figure 7.5b-2), the estimation function of the CDOB maintains a small estimation error even in the presence of disturbances.

7.3 Control with Communication DOB Under Disturbance

The simulation results in Section 7.2 show that the CDOB is very effective in reducing communication time fluctuations under conditions where disturbances can be ignored but problems arise when disturbances cannot be ignored. We can perform a disturbance suppression control if the terminal-side mechatronics system includes a robust controller to solve this problem. In that case, the mechatronics control system of the terminal side is treated as $G(s)$. Then, it is available to construct a control system with a CDOB from the operator's side. Figure 7.6 shows the basic configuration diagram. However, in this configuration example, the time delay is only on the input side, and the design of the mechatronics system-side controller $C_m(s)$ can be left to the designer's specialty. The mechatronics system-side control system $G(s)$ can be an acceleration control system, a velocity control system, a position control system, or any other system. The following is an example of a system expected.

Example 7.3.1 (Example with modern control.) We introduce an example in which the mechatronics system controller $C_m(s)$ is a position servo system based on the modern control method. Here, the apparent $G(s)$ is approximated by a general second-order delay system, and the controller on the operator side is also a position servo system. The Simulink model is shown in Figure 7.7. Additionally, the position waveforms with and without CDOB are shown in Figure 7.8a,b, respectively. The corresponding program list is shown in List 7.2.

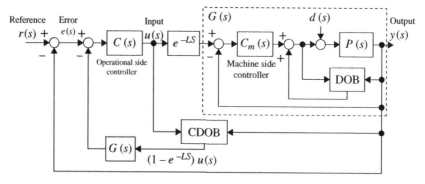

Figure 7.6 Configuration of control system with CDOB under disturbances.

List 7.2:   CDOB-based servo control with DOB.

```
1   %% Physical parameters and PD controller settings
2   m=2; % Mass[kg]
3   L=0.2;L_est=L*0.25;% Wasteland time L and estimate L_est
4   w=2;c=1; % w:Angular frequency of position reference[rad/s]
5
6   %% Mechatronics system–side DOB design and servo system design
7   % in state–space representation
8   Ap=[0,1;0,-c/m];Bp=[0;1/m];Cp=[1,0];
9   Apbar=[Ap,-Bp;zeros(1,2),0];Cpbar=[Cp,0];
10  pole_pob=[-10,-12,-14];
11  Hptemp=place(Apbar',Cpbar',pole_pob);Hp=Hptemp';
12  Hp1=Hp(1:2);Hp2=Hp(3); % DOB gain on mechatronics system side
13  %
14  Ap_tilde=[Ap,zeros(2,1);-Cp,0];Bp_tilde=[Bp;0];
15  pole=[-10,-20,-30];F_tilde=place(Ap_tilde,Bp_tilde,pole);
16  Fpp=F_tilde(1:2);Kii=-F_tilde(3); % Servo gains of slave side
17
18  %% CDOB design for the operator in state–space representation
19  zeta=1;wn=10;
20  A=[0 1;-wn^2,-2*zeta*wn];
21  B=[0; 1];C=[wn^2,0];
22  Abar=[A,-B;0,0,0];Cbar=[C,0];
23  pole=[-10,-12, -13]*10;
24  Htemp=place(Abar',Cbar',pole);H=Htemp';
25  H1=H(1:2);H2=H(3);
26  %
27  AA=[A,zeros(2,1);-C,0];BB=[B;0];
28  pole=[-2,-3,-4]*2;FF_tilde=place(AA,BB,pole);
29  FF=FF_tilde(1:2);KKi=-FF_tilde(3); % Servo gains of master side
30
31  %% Simulation
32  tend=20; % Simulation time
33  t_dis=tend*2/3;dis=1.5; % Disturbance generation time and magnitude
34  open_system('sim_Figure_7_7_and_7_8_servo_with_CDOB');
35  z=sim('sim_Figure_7_7_and_7_8_servo_with_CDOB');
```

The internal structure of the DOB in Figure 7.7 is the same as in Chapters 2 and 4. Moreover, the CDOB blocks in Figure 7.7 is the same as in Section 7.2. The F_{pp} and FF are the state feedback gain matrices for the mechatronics system and the operator, respectively, and K_{ii} and KK_i are the respective integral gains. The servo control system model is $\frac{\omega_n^2}{s^2+2\zeta\omega_n s+\omega_n^2}$ with resonance angular frequency ω_n and damping constant ζ approximating $G(s)$, but has a slight shift from the actual $G(s)$

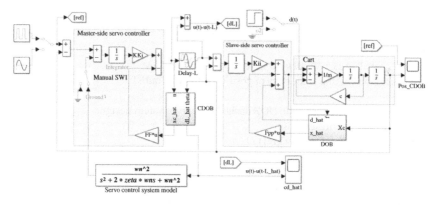

Figure 7.7 Simulink model of control system with CDOB in the presence of disturbance.

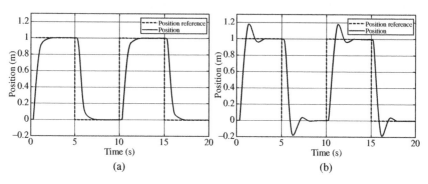

Figure 7.8 Simulation results with time delay and disturbance. (a) Position waveform (with CDOB) and (b) position waveform (without CDOB).

property intentionally, to evaluate robustness. Figure 7.8 (a) shows the position control result when the manual switch 1 of Figure 7.7 is set to the right with feedback from the CDOB. Oppositely, Figure 7.8 (b) shows the position control result when the manual switch 1 of Figure 7.7 is set to the left without feedback from the CDOB. In the presence of disturbance, oscillation is observed in Figure 7.8b, while good control is achieved in (a).

What if the mechatronics system cannot be equipped with its own controller? In particular, a method that can compensate for model errors has been reported [4], but is not included in this chapter.

References

1 O. Journal M. Smith: A control to overcome dead time, ISA Journal, Vol. 6, No. 2, 28–33, 1959.

2 Keiji Watanabe, Yasunori Ishiyama, Masami Ito: Modified Smith predictor control for multivariable systems with delays and unmeasurable step disturbances, International Journal of Control, Vol. 37, No. 5, 959–973, 1983.

3 Kenji Natori, Kouhei Ohnishi: A design method of communication disturbance observer for time-delay compensation, taking the dynamic property of network disturbance into account, IEEE Transactions of Industrial Electronics, Vol. 55, No. 5, 2152–2168, 2008.

4 Naoya Uematsu, Tooru Suhara, Yutaka Uchimura: Mode-error feedback for systems with time-varying delay based on mixed H_2/H_∞ control, IEEE Journal of Industry Applications, Vol. 5, No. 3, 245–252, 2016.

8

Multirate Disturbance Observer

Multirate control is an effective control method when using information acquired from devices with relatively long processing times, such as image sensors and LiDAR sensors, along with devices that require short control cycles, such as motor control. This chapter introduces two methods of multirate disturbance estimation.

8.1 Multirate System Modeling

A control system with multiple control cycles is called a **multirate control system** [1–4]. There is a multirate control when the control period T_u on the input side and the observation period T_y on the observation side are different, both when $T_u < T_y$ and when $T_u \geq T_y$. This book focuses on the former case called **input multiplexing**. It considers an observer such that the control input $u(t)$ is updated every control period $T_u (= T_y/N)$, dividing the observation period T_y by N.

In Figure 8.1, representing the concept, $x\left(i + \frac{k}{N}\right)$ ($k = 0, 1, 2, \ldots, N-1$) denotes $x(t)$ at $t = \left(i + \frac{k}{N}\right)T_y$. Almost the same for $u(t)$, but it is assumed to be the zeroth order held during T_u.[1]

Here, the continuous system's equation of state $\dot{x}(t) = Ax(t) + Bu(t)$ is digitized with the control period T_u and is represented by Equation (8.1). This section uses the subscripts s and l (short and long, respectively), depending on the two types of control cycles.

$$x\left(i + \frac{k+1}{N}\right) = A_s x\left(i + \frac{k}{N}\right) + B_s u\left(i + \frac{k}{N}\right) - B_s d\left(i + \frac{k}{N}\right) \tag{8.1}$$

1 This book uses the control period T_y as the basic unit, and the $t = (i + kN)T_y$ represents the time at each sample point of T_u. Meanwhile, we can get the equation of $\left(i + \frac{k}{N}\right)T_y = (Ni + k)T_u$ since $T_y = N \cdot T_u$ from the above definition. So, we can also express $x(Ni + k)$ instead of $x\left(i + \frac{k}{N}\right)$, with T_u as the unit. The choice depends on the designer.

Disturbance Observer for Advanced Motion Control with MATLAB/Simulink, First Edition. Akira Shimada.
© 2023 The Institute of Electrical and Electronics Engineers, Inc. Published 2023 by John Wiley & Sons, Inc.
Companion website: www.wiley.com/go/disturbanceobserver

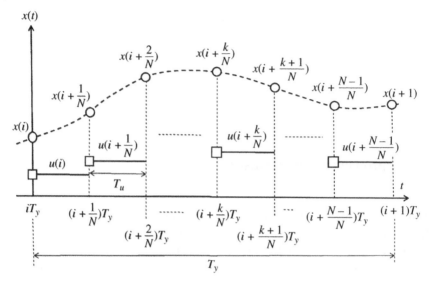

Figure 8.1 Conceptual diagram of state variables and inputs in multirate.

where $k = 0, 1, \ldots, N - 1$ and $A_s = exp(AT_u)$, $B_s = \int_0^{T_u} exp(A\sigma)d\sigma B$.

We assume the disturbance $d\left(i + \frac{k+1}{N}\right) = d\left(i + \frac{k}{N}\right)$ and the extended state

variable $\bar{x}\left(i + \frac{k}{N}\right) = \begin{bmatrix} x\left(i + \frac{k}{N}\right) \\ d\left(i + \frac{k}{N}\right) \end{bmatrix}$ to create the extended system equation (8.12):

$$
\begin{bmatrix} x\left(i + \frac{k+1}{N}\right) \\ d\left(i + \frac{k+1}{N}\right) \end{bmatrix} = \begin{bmatrix} A_s & -B_s \\ 0_{m \times n} & I_m \end{bmatrix} \begin{bmatrix} x\left(i + \frac{k}{N}\right) \\ d\left(i + \frac{k}{N}\right) \end{bmatrix}
$$
$$
+ \begin{bmatrix} B_s \\ 0 \end{bmatrix} u\left(i + \frac{k}{N}\right) \tag{8.2}
$$

Let the coefficient matrix of the first term on the right-hand side in Equation (8.2) be $\bar{A}_s$, and the coefficient matrix of the second term be $\bar{B}_s$. Then, create a state equation with the multiplexed input in which the control period T_y is divided by N:

$$
\bar{x}\left(i + \frac{1}{N}\right) = \bar{A}_s \bar{x}(i) + \bar{B}_s u(i) \tag{8.3}
$$
$$
\bar{x}\left(i + \frac{2}{N}\right) = \bar{A}_s \bar{x}\left(i + \frac{1}{N}\right) + \bar{B}_s u\left(i + \frac{1}{N}\right)
$$
$$
= \bar{A}_s \{\bar{A}_s \bar{x}(i) + \bar{B}_s u(i)\} + \bar{B}_s u\left(i + \frac{1}{N}\right)
$$
$$
= \bar{A}_s^2 x[i] + \bar{A}_s \bar{B}_s u(i) + \bar{B}_s u\left(i + \frac{1}{N}\right)
$$

$$= \bar{A}_s^2 x(i) + [\bar{A}_s \overline{B}_s, \overline{B}_s] \begin{bmatrix} u(i) \\ u\left(i + \frac{1}{N}\right) \end{bmatrix}$$

$$\vdots$$

$$\bar{x}(i+1) = \bar{A}_s^N x(i) + [\bar{A}_s^{N-1} \overline{B}_s, \dots, \overline{B}_s] \begin{bmatrix} u(i) \\ u\left(i + \frac{1}{N}\right) \\ \vdots \\ u\left(i + \frac{N-1}{N}\right) \end{bmatrix}$$

$$= \bar{A}_l x(i) + \overline{B}_l \bar{u}(i) \tag{8.4}$$

It means that $\bar{A}_l = \bar{A}_s^N \in R^{(n+m)\times(n+m)}$, and $\overline{B}_l = [\bar{A}_s^{N-1}\overline{B}_s, \dots, \overline{B}_s] \in R^{(n+m)\times mN}$. The output is $y(i) = \overline{C}_l \bar{x}(i)$, where $\overline{C}_l = [C_s, 0_{l\times m}] \in R^{l\times(n+m)}$. The transformation operation that cuts the input between T_y into N pieces and arranges them in vector form, as in $\bar{u}(i)$ in Equation (8.4), is called "**lifting.**"

8.2 Multirate Disturbance Observer (Method 1)

A **multirate disturbance observer** can be designed if the above $(\bar{A}_l, \overline{C}_l)$ is observable. Two types of design methods are presented in this book.

8.2.1 Disturbance Observer Design (Method 1)

Assuming $(\bar{A}_l, \overline{C}_l)$ is observable, design a multirate observer of Equations (8.5) and (8.6) from Equation (8.4) and $y(i) = \overline{C}_l \bar{x}(i)$:

$$\hat{\bar{x}}(i+1) = \bar{A}_l \hat{\bar{x}}(i) + \overline{B}_l \bar{u}(i) + H(y(i) - \hat{y}(i)) \tag{8.5}$$

$$\hat{y}(i) = \overline{C}_l \hat{\bar{x}}(i) \tag{8.6}$$

$H = [H_1, H_2]^T$ is the observer gain designed using the pole assignment method, LQ[2] design method, etc.

Here again, expressing the observer by element yields the equation

$$\begin{bmatrix} \hat{x}(i+1) \\ \hat{d}(i+1) \end{bmatrix} = \begin{bmatrix} A_l & -B_l \\ 0_{m\times n} & I_m \end{bmatrix} \begin{bmatrix} \hat{x}(i) \\ \hat{d}(i) \end{bmatrix} + \begin{bmatrix} B_l \\ 0_{m\times N} \end{bmatrix} \bar{u}(i)$$

$$+ \begin{bmatrix} H_1 \\ H_2 \end{bmatrix} \{y(i) - \hat{y}(i)\} \tag{8.7}$$

$$\hat{y}(i) = [C_l, 0] \begin{bmatrix} \hat{x}(i) \\ \hat{d}(i) \end{bmatrix} \tag{8.8}$$

However, $\overline{B}_l = [B_l, 0_{m\times N}]^T$, and $-B_l$ is the upper right element of $\bar{A}_l$.

---

2 See Appendix A.3.7 about LQ control.

8.2.2 Controller Design Using Multirate Observer (Method 1)

This observer estimates the extended state variable $\bar{x}(i) = [x^T(i), d^T(i)]^T$ for each control cycle T_y using $\bar{u}(i)$ with lifting inputs for each control cycle T_u. Therefore, if we try generating control inputs for each T_u by feeding back the estimated values, the dimensions and timing will not match. Therefore, we use the estimated $\hat{x}[i]$ and the multiplexed input vector $\bar{u}[i]$ to calculate the state variable vector at each T_u,

$$x_g[i] = \begin{bmatrix} \hat{x}(i) \\ \hat{x}\left(i + \frac{1}{N}\right) \\ \vdots \\ \hat{x}\left(i + \frac{N-1}{N}\right) \end{bmatrix} = A_g\hat{x}(i) + B_g\bar{u}(i) - B_{g2}\hat{d}(i) \tag{8.9}$$

where $A_g = \begin{bmatrix} I_{n \times n} \\ A_s \\ \vdots \\ A_s^{N-1} \end{bmatrix}$, $B_g = \begin{bmatrix} 0_{n \times m} & 0_{n \times m} & \cdots & 0_{n \times m} \\ B_s & 0_{n \times m} & \cdots & \vdots \\ \vdots & \ddots & \ddots & \vdots \\ A_s^{N-2}B_s & \cdots & B_s & 0_{n \times m} \end{bmatrix}$

$B_{g2} = \begin{bmatrix} 0_{m \times n}^T & B_s^T & \cdots & (A_s^{N-2}B_s)^T \end{bmatrix}^T$

Assuming the state feedback as an example, we need a feedback gain F such that $u = -Fx$, but since we have estimated the state $x_g(i)$ consisting of N elements for each T_u, we can use the pole assignment or LQ control method. After obtaining F using $F_l = diag(F, F, \dots, F)$ as the control input, we get Equation (8.10):

$$\bar{u}(i) = -F_l x_g(i) = -F_l\{A_g\hat{x}(i) + B_g\bar{u}(i) - B_{g2}\hat{d}(i)\} \tag{8.10}$$

However, $\bar{u}[i]$ is on the left and right, so we tidy up and obtain Equation (8.11):

$$\bar{u}[i] = -(I + F_l B_g)^{-1}F_l\{A_g\hat{x}(i) - B_{g2}\hat{d}(i)\} \in R^N \tag{8.11}$$

The control input $\bar{u}(i)$ is a vector determined for each T_y with N components, which cannot be used as a control input for each T_u. Therefore, we define a function called the **multirate hold** (H_M). It is a function that switches N pairs of components of the control input vector sequentially during the control cycle

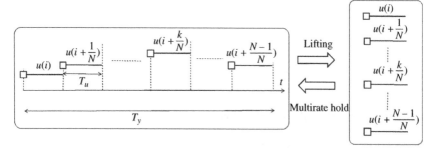

Figure 8.2 Conceptual diagram of input conversion in multirate control.

T_y at every control cycle T_u. That is, the control input $\bar{u}(i)$ is distributed to $u\left(i + \dfrac{k}{N}\right)$ $(k = 1, 2, \ldots, N - 1)$ through multirate hold H_M and input to the control plant. See Figure 8.2.

Example 8.2.1 An example program with $N = 4$ is shown in List 8.1, the Simulink model and the settings in Figure 8.3, and a simulation result in Figure 8.4.

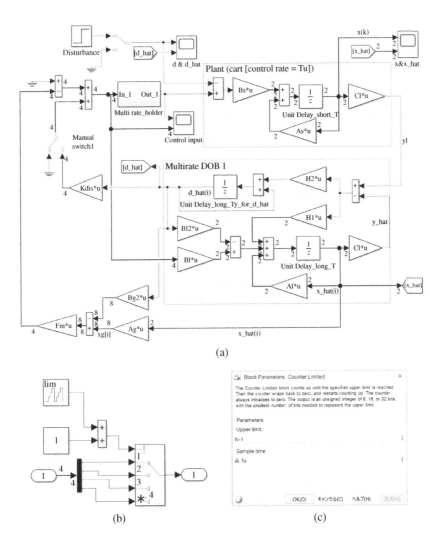

(a)

(b)   (c)

Figure 8.3 Simulink model of control system with multirate DOB (Method 1). (a) The entire Simulink model is available, (b) multirate holder example, and (c) counter limited setting.

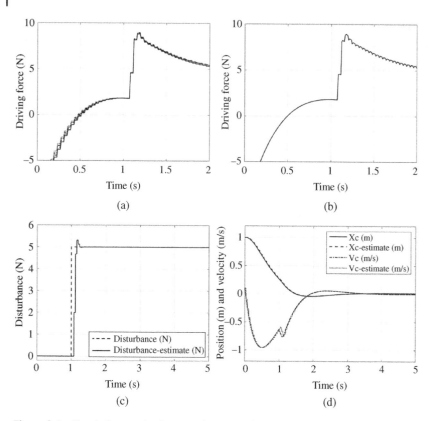

Figure 8.4 Simulation results for control system with multirate disturbance observer (DOB) (Method 1). (a) Multiple input waveform at $N = 4$, (b) input waveform for each T_u, (c) disturbance waveform, and (d) state waveform.

List 8.1: Control system with multirate disturbance observer. Source: Adapted from [1].

```
1    %% Physical parameters
2    mc=2.0;c=0;dis=5; % Mass and viscosity coefficient
3    % and magnitude of disturbance
4
5    %% Parameter design of continuous system
6    A=[0 1;0,-c/mc];B=[0;1/mc];C=[1,0];D=0;Cl=C;
7    n=2;m=1;l=1; % Generally,[n,m]=size(B);[l,~]=size(C);Cl=C;
8
9    %% Parameter design for digital system
10   N=4;Tu=0.01;Ty=Tu*N; % As an example,Ty=4*Tu
11   [As,Bs,Cs,Ds]=c2dm(A,B,C,D,Tu);[~,Bl2,~,~]=c2dm(A,B,C,D,Ty);
12
```

```
13   As_bar=[As,-Bs;zeros(1,2),1];Bs_bar=[Bs;0];Cl_bar=[Cs,0];
14   Al=As^N;Al_bar=As_bar^N; % A matrix with multiple sample
15
16   Bl=[];for i=1:N Bl=[Bl,As^(N-i)*Bs];end;% B1 matrix with multiple sample
17   Bl_bar=[];for i=1:N Bl_bar=[Bl_bar,As_bar^(N-i)*Bs_bar];end;
18   % Bl:Multiplexing of Bs, Bl_bar:Multiplexing of Bs_bar
19
20   Uo=[Cl_bar;Cl_bar*Al_bar;Cl_bar*Al_bar^2];rank_Uo=rank(Uo)
21
22   %% Multi-rate disturbance observer design
23   if rank_Uo==n+1
24       fprintf('Observable\n');
25       Q_ob=diag([1e1,1e3,1e8]);R_ob=1;
26       Htemp=dlqr(Al_bar',Cl_bar',Q_ob,R_ob);H=Htemp';
27       H1=H(1:2);H2=H(3);
28   else fprintf('Unobservable\n');
29   end;
30
31   %% Controller Design
32   Q=diag([100,10]);R=1;F=dlqr(As,Bs,Q,R);
33   Fl=blkdiag(F,F,F,F); % Multiple feedback gain
34   Ag=[eye(n);As;As^2;As^3];
35   Bg=[zeros(2,4);
36       Bs,zeros(2,3);
37       As*Bs,Bs,zeros(2,2);
38       As^2*Bs,As*Bs,Bs,zeros(2,1)];
39   Fm=inv(eye(4)+Fl*Bg)*Fl;
40
41   Bg2=[zeros(2,1);Bs;As*Bs;As^2*Bs];
42   Kdis=ones(4,1);
43
44   %% Simulation
45   x0=[1.0;0.1];x0_ob=[0;0];tend=5.0;
46   open_system('sim_Figure_8_3_and_8_4_multirate_DOB1');
47   set_param('sim_Figure_8_3_and_8_4_multirate_DOB1','WideLines','on');
48   set_param('sim_Figure_8_3_and_8_4_multirate_DOB1','ShowLineDimensions','on');
49   z=sim('sim_Figure_8_3_and_8_4_multirate_DOB1');
```

The K_{dis} in the program is the gain to line up $N(= T_y/T_u)$ of the disturbance estimate that does not change between T_y and makes the output of K_{dis} return to the input side of the multirate hold with the positive feedback. Figure 8.4a shows the overlap of four multiple input values computed between T_y due to $N = 4$, (b) shows the input waveforms connected by the multirate hold H_M, which is switched every T_u, and Figure 8.4c shows the disturbance estimation, while (d) shows the position and velocity waveforms. The estimated values follow the actual position and velocity.

8.3 Multirate Disturbance Observer (Method 2)

This section introduces a simpler method for designing multirate disturbance observers [5, 6]. The design method makes use of a **current state observer**. This current state observer performs estimation in two steps as the Kalman filter: a priori and a posteriori estimation. However, this observer uses a method in which the prediction computation is iteratively updated every T_u and modified every T_y using the observer gain L. The superscript $\bar{x}$ denotes the extended system $[x^T, d^T]^T$, $\tilde{x}$ denotes the prior estimate, and $\hat{x}$ the posterior estimate. Thus, the prior estimate of the extended system is $\tilde{\bar{x}}$, and the posterior estimate of the extended system is denoted by $\hat{\bar{x}}$. As a priori estimate, we assume that for every T_u, the result of Equation (8.12) is delayed by one sample:

$$
\begin{aligned}
\tilde{\bar{x}}\left(i + \frac{k}{N}\right) &= \begin{bmatrix} \tilde{x}\left\{(i-1) + \frac{k}{N}\right\} \\ \tilde{d}\left\{(i-1) + \frac{k}{N}\right\} \end{bmatrix} \\
&= \bar{A}_s \hat{\bar{x}}\left(i + \frac{k-1}{N}\right) + \bar{B}_s u\left(i + \frac{k-1}{N}\right) \\
&= \begin{bmatrix} A_s & -B_s \\ 0_{m \times n} & I_m \end{bmatrix} \begin{bmatrix} \hat{x}\left(i + \frac{k-1}{N}\right) \\ \hat{d}\left(i + \frac{k-1}{N}\right) \end{bmatrix} \\
&\quad + \begin{bmatrix} B_s \\ 0 \end{bmatrix} u\left(i + \frac{k-1}{N}\right)
\end{aligned}
\tag{8.12}
$$

As a posteriori estimation, we compute Equation (8.13):

$$
\begin{aligned}
\hat{\bar{x}}\left(i + \frac{k}{N}\right) &= \begin{bmatrix} \hat{x}\left(i + \frac{k}{N}\right) \\ \hat{d}\left(i + \frac{k}{N}\right) \end{bmatrix} \\
&= \begin{cases} \tilde{\bar{x}}(i) + L \cdot e\left(i + \frac{k}{N}\right) & (k = 0) \\ \tilde{\bar{x}}\left(i + \frac{k}{N}\right) & (k = 1, 2, \ldots, N-1) \end{cases} \\
&= \begin{cases} \begin{bmatrix} \tilde{x}\left(i + \frac{k}{N}\right) \\ \tilde{d}\left(i + \frac{k}{N}\right) \end{bmatrix} + \begin{bmatrix} L_1 \\ L_2 \end{bmatrix} e\left(i + \frac{k}{N}\right) & (k = 0) \\ \begin{bmatrix} \tilde{x}\left(i + \frac{k}{N}\right) \\ \tilde{d}\left(i + \frac{k}{N}\right) \end{bmatrix} & (k = 1, 2, \ldots, N-1) \end{cases}
\end{aligned}
\tag{8.13}
$$

where $e\left(i + \frac{k}{N}\right) = y\left(i + \frac{k}{N}\right) - \hat{y}\left(i + \frac{k}{N}\right)$.[3] If $e_x(i) = \hat{\bar{x}}(i) - \bar{x}(i)$ at $k = 0$, that is the state estimation error for each T_y, then

$$
\begin{aligned}
e_x(i) &= \hat{\bar{x}}(i) - \bar{x}(i) = \{\tilde{\bar{x}}(i) + L\{y(i) - \hat{y}(i)\} - \bar{x}(i) \\
&= \{\tilde{\bar{x}}(i) + L\{\overline{C}_l\bar{x}(i) - \overline{C}_l\tilde{\bar{x}}(i)\} - \bar{x}(i) \\
&= (I_{n+m} - L\overline{C}_l)\tilde{\bar{x}}(i) - (I_{n+m} - L\overline{C}_l)\bar{x}(i) = (I_{n+m} - L\overline{C}_l)\{\tilde{\bar{x}}(i) - \bar{x}(i)\} \\
&= (I_{n+m} - L\overline{C}_l)[\{\bar{A}_l\hat{\bar{x}}(i-1) + \overline{B}_l\bar{u}(i-1)\} - \{\bar{A}_l\bar{x}(i-1) + \overline{B}_l\bar{u}(i-1)\}] \\
&= (\bar{A}_l - L\overline{C}_l\bar{A}_l)e_x(i-1)
\end{aligned}
\tag{8.14}
$$

From this result, we design L so that $\bar{A}_l - L\overline{C}_l\bar{A}_l$ has a stable pole.

Example 8.3.1 An example program for Method 2 with $N = 4$ is shown in List 8.2, a Simulink model is shown in Figure 8.5, and Figure 8.6 shows a simulation example.

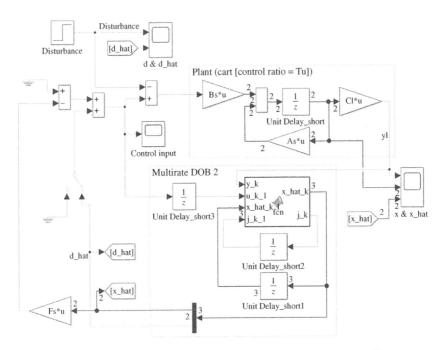

Figure 8.5 Simulink model of control system with multirate DOB (Method 2).

---

3 In the choice in Equation (8.13), when $k = 0$, $i + \frac{k}{N} = i$, so the brackets are (i), but we left $+\frac{k}{N}$ to show that it is "when $k = 0$" after $\frac{k}{N}$ is present.

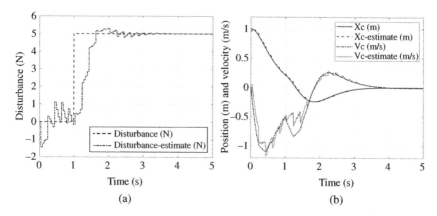

Figure 8.6 Simulation results of multirate DOB-combined control system (Method 2). (a) Disturbance waveform and (b) state waveform.

List 8.2:   Multirate disturbance observer-combined control system 2.

```
1   %% Physical paremeters
2   mc=2.0;c=0;dis=5; % Mass and viscosity coefficient
3   % and magnitude of disturbance
4
5   %% Parameter design of continuous system
6   A=[0 1;0,-c/mc];B=[0;1/mc];C=[1,0];D=0;Cl=C;
7   n=2;m=1;l=1; % Generally,[n,m]=size(B);[l,~]=size(C);Cl=C;
8   dis=5;
9
10  %% Parameter design for digital system
11  N=4;Tu=0.01;Ty=Tu*N; % As an example,Ty=4*Tu
12  [As,Bs,Cs,Ds]=c2dm(A,B,C,D,Tu);[~,Bl2,~,~]=c2dm(A,B,C,D,Ty);
13  As_bar=[As,-Bs;zeros(1,2),1];Bs_bar=[Bs;0];
14  Cs_bar=[Cs,0];
15
16  %% Disturbance observer design
17  Al_bar=As_bar^N;Cl_bar=Cs_bar;
18  Q_ob=diag([1e1,1e1,2e4]);R_ob=1;
19  Ltemp=dlqr(Al_bar',(Cl_bar*Al_bar)',Q_ob,R_ob);L=Ltemp';
20
21  %% Controller design
22  Q=diag([100,10]);R=1;Fs=dlqr(As,Bs,Q,R)
23
24  %% Initial values
25  x0=[1.0;0.01];x0_ob=[0.96;0.005;0];
26  tend=5.0;
27  open_system('sim_Figure_8_5_and_8_6_multirate_DOB2');
```

```
28   set_param('sim_Figure_8_5_and_8_6_multirate_DOB2','WideLines','on');
29   set_param('sim_Figure_8_5_and_8_6_multirate_DOB2'
```

List 8.3:   Multirate DOB2 MATLAB function.

```
1   function [x_hat_k,j_k] = fcn(y_k,u_k_1,x_hat_k_1,j_k_1,As_bar,Bs_bar,Cl_bar,L,N)
2   %.
3   x_bar_k=As_bar*x_hat_k_1+Bs_bar*u_k_1;
4   if j_k_1>N-1
5   x_hat_k =x_bar_k+L*(y_k-Cl_bar*x_bar_k); j_k=0;
6   else x_hat_k=x_bar_k; j_k=j_k_1+1;end;
```

List 8.3 is the function program programmed in the MATLAB function defined in Figure 8.5.

Considering Figure 8.6a, the disturbance estimate can track the real disturbance, but the waveform appears a bit violently. Meanwhile, the state estimation considering the disturbance shown in Figure 8.6b is successful, and the control input can converge to 0.[4]

References

1 Hiroshi Fujimoto, Yoichi Hori, Atsuo Kawamura: Perfect tracking control based on multirate feedfoward control with generalized sampling periods, Transactions on IEEE Industrial Electronics, Vol. 48, No. 3, 636–644, 2001.

2 Takeyori Hara, Masayoshi Tomizuka: Multi-Rate Controller for Hard Disc Drives with Redesign of State Estimator, Proceedings of the 1998 American Control Conference, 3033–3037, 1998.

3 Akira Shimada, Tsuyoshi Takeda: Multirate Feedforward Control of Robot Manipulators, 2007 IEEE/RSJ International Conference on Intelligent Robots and Systems (IROS2007), ThD6.5, 4083–4088, 2007.

4 Akira Shimada, Atsushi Kusakari: Multirate disturbance observer based controller design on automatic controltive visions using parallel mechanisms, Transactions of SICE, Vol. 44, No. 1, 18–25, 2008 (In Japanese).

5 Hiroshi Fujimoto, Yoichi Hori: Visual Servoing Based on Intersample Disturbance Rejection by Multirate Sampling Control, Proceedings of the 40th IEEE Conference on Decision and Control, TuA11-6, 334–339, 2001.

6 Takashi Yamaguchi, Mitsuo Hirata, Chee Khiang Pang: High-Speed Precision Motion Control, CRC Press, 2011.

---

4 We run the simulation with a short control cycle T_u although the observer gain, L, is designed with a long control cycle T_y. Actually, the calculation using L in the MATLAB function is performed at every T_y.

9

Model Predictive Control with DOB

An optimal control method that considers the constraints is **model predictive control** (MPC).[1] This chapter gives an overview of MPC, citing representative references [2, 3], and the method for combining it with a disturbance observer is introduced [4].

9.1 Model Predictive Control (MPC)

9.1.1 Overview of MPC

The model predictive control is a control method that explicitly uses optimal programming and is characterized by solving a finite-time optimal control problem under constraints and repeating the process using only the first control input values at each control cycle. Unlike conventional control methods, stability cannot be guaranteed as in the LQ[2] control or pole assignment method. However, the advantage of this method is that the control system can directly incorporate the constraint conditions. Introducing the MPC system with a disturbance observer is that the model predictive control requires finite-time predictions using all the control plant states, but not all state quantities can be observed. In addition, unknown disturbances may have a significant effect. Therefore, designing a combined system with a disturbance observer is practical.

Figure 9.1a is a typical MPC block diagram. The optimal controller is the optimizer, which outputs $\Delta u(k)$ of the change in the control input. The actual control input $u(k)$ is its integral value, but it differs from the so-called type 1

---

1 We need the basic knowledge of optimization to learn MPC theory [1] . However, since this is not a specialized book on optimal programming, it contains only a brief explanation and MATLAB's fmincon function. We describe a simple example using the fmincon function in Appendix A.6.
2 See Appendix A.3.7 about LQ control.

Disturbance Observer for Advanced Motion Control with MATLAB/Simulink, First Edition. Akira Shimada.
© 2023 The Institute of Electrical and Electronics Engineers, Inc. Published 2023 by John Wiley & Sons, Inc.
Companion website: www.wiley.com/go/disturbanceobserver

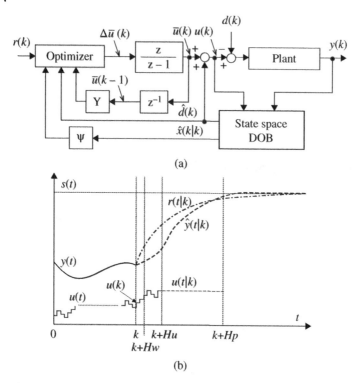

Figure 9.1 Model predictive control system. (a) Block diagram of model predictive control and (b) waveform of model predictive control.

servo system because it does not integrate the deviation as it is. However, since it requires information on all state variables, it is commonly used in conjunction with state estimation functions such as observers and Kalman filters. This seems to work particularly well combined with disturbance observers.

Figure 9.1b illustrates the basic characteristics of the waveforms [2]. $S(t)$ is the setpoint trajectory, $y(t)$ is the observed output, $u(t)$ is the control input, and $r(t|k)$ is the reference trajectory, which is the modified value toward $S(t)$ from the point $t = kT$. However, $S(t)$ and $r(t|k)$ may sometimes not be separated. The value $(t|k)$ represents the predicted time value after the $t = kT$ point in time. Similarly, $(k + 1|k), (k + 2|k), \ldots$ denote the predicted values of $t = (k + 1)T, t = (k + 2)T, \ldots$ from the time $t = kT$. Moreover, $H_w, H_u,$ and H_p are said to be the **"window parameters," "control horizon,"** and **"prediction horizon,"** and the designer decides them.

Specifically, find the input $\Delta u(k)$ that minimizes the objective function $V(k) = \sum_{i=H_w}^{H_p} \|\hat{z}(k+i|k) - r(k+i|k)\|_{Q(i)}^2 + \sum_{i=0}^{H_u-1} \|\Delta \hat{u}(k+i|k)\|_{R(i)}^2$. Here, $\hat{z}(k+i)$ is the predicted value of the output we wish to control. The main feature of MPC is to find the time series data sequence of optimal control input changes in the finite time $\Delta U(k) = [\Delta \hat{u}(k|k), \ldots, \Delta \hat{u}(k+H_u-1|k)]^T$; $\Delta u(k)$ repeatedly uses the first value as a control input.

9.1.2 Formulation and Objective Function for the MPC Design

Let us write the equation of state as in Equation (9.3). The notation $(k+i|k)$ denotes the predicted $\hat{x}, \hat{u}$ of x or u at $k+i$ viewed from the k point in time. The **prediction horizon** H_p is used to consider forecasts from the time $t = kT$ to the time $t = (k+H_p)T$:

$$\hat{x}(k+1|k) = A_d\hat{x}(k) + B_d\hat{u}(k|k)$$

$$\hat{x}(k+2|k) = A_d\hat{x}(k+1|k) + B_d\hat{u}(k+1|k)$$

$$= A_d\{A_d\hat{x}(k) + B_d\hat{u}(k|k)\} + B_d\hat{u}(k+1|k)$$

$$= A_d^2\hat{x}(k) + A_dB_d\hat{u}(k|k) + B_d\hat{u}(k+1|k)$$

$$\vdots$$

$$\hat{x}(k+H_p|k) = A_d^{H_p}\hat{x}(k)$$

$$+ A_d^{H_p-1}B_d\hat{u}(k|k) + A_d^{H_p-2}B_d\hat{u}(k+1|k)$$

$$+ \cdots + B_d\hat{u}(k+H_p-1|k) \tag{9.1}$$

Next, the predicted value $\hat{u}$ of the control input u from $t = kT$ to $t = (k+H_u-1)T$ is expressed in Equation (9.2). However, the control horizon H_u is assumed to satisfy $H_u \leq H_p - 1$:

$$\hat{u}(k|k) = \Delta\hat{u}(k|k) + u(k-1)$$

$$\hat{u}(k+1|k) = \Delta\hat{u}(k+1|k) + \hat{u}(k|k)$$

$$= \Delta\hat{u}(k+1|k) + \{\Delta\hat{u}(k|k) + u(k-1)\}$$

$$\vdots$$

$$\hat{u}(k+H_u-1|k) = \Delta\hat{u}(k+H_u-1|k) + \cdots + \Delta\hat{u}(k|k) + u(k-1) \tag{9.2}$$

The $\hat{u}$ is not changed, and $\Delta\hat{u} = 0$ after $t = (k+H_u+1)T$, as in Figure 9.1b.

Next, consider the predicted value $\hat{y}(t|k)$ of the observed output $y(k)$ and the predicted value $\hat{z}(t|k)$ of the control quantity $z(k)$. The physical quantity z we want to control need not necessarily be the observed output. Therefore, let $z(k) = C_z x(k)$; that is, let $\hat{z}(t|k) = C_z\hat{x}(t|k)$ be the predicted value.

Combining Equations (A.33) and (9.2), and considering the above, we obtain Equation (9.3):

$$
\begin{bmatrix} \hat{z}(k+1|k) \\ \hat{z}(k+2|k) \\ \vdots \\ \hat{z}(k+H_w|k) \\ \vdots \\ \hat{z}(k+H_p|k) \end{bmatrix} = \begin{bmatrix} C_z A_d \\ C_z A_d^2 \\ \vdots \\ C_z A_d^{H_w} \\ \vdots \\ C_z A_d^{H_p} \end{bmatrix} x(k) + \begin{bmatrix} C_z B_d \\ C_z A_d B_d \\ \vdots \\ C_z \sum_{i=0}^{H_w-1} A_d^i B_d \\ \vdots \\ C_z \sum_{i=0}^{H_p-1} A_d^i B_d \end{bmatrix} u(k-1)
$$

$$
+ \begin{bmatrix} C_z B_d & 0 & \cdots & & 0 \\ C_z A_d B_d & C_z B_d & 0 \cdots & & 0 \\ \vdots & & \ddots & & \vdots \\ C_z \sum_{i=0}^{H_w-1} A_d^i B_d & \cdots & & & C_z B_d \\ \vdots & & & & \vdots \\ C_z \sum_{i=0}^{H_p-1} A_d^i B_d & \cdots & & \cdots & C_z \sum_{i=0}^{H_p-H_u} A_d^i B_d \end{bmatrix} \Delta U(k) \quad (9.3)
$$

where
$$
\Delta U(k) = \begin{bmatrix} \Delta u(k|k) \\ \Delta \hat{u}(k+1|k) \\ \vdots \\ \Delta \hat{u}(k+H_u-1|k) \end{bmatrix}.^3
$$

Next, to express the objective function, let $Z(k)$ denote the part from line H_w to line H_p of the left-hand side in Equation (9.3), and let $\Psi(k)$ denote the part from line H_w to line H_p of the first term on the right-hand side. Υ denotes the same part of the coefficient matrix in the second term on the right-hand side. Finally, let Θ denote the same part of the coefficient matrix of the third term. $Z(k) = \begin{bmatrix} \hat{z}(k+H_w|k) \\ \vdots \\ \hat{z}(k+H_p|k) \end{bmatrix}$,

$$
\Psi = \begin{bmatrix} C_z A_d^{H_w} \\ \vdots \\ C_z A_d^{H_p} \end{bmatrix}, \Upsilon = \begin{bmatrix} C_z \sum_{i=0}^{H_w-1} A_d^i B_d \\ \vdots \\ C_z \sum_{i=0}^{H_p-1} A_d^i B_d \end{bmatrix}.
$$

Finally, Let's define Θ as the followings:

$$
\Theta = \begin{bmatrix} C_z \sum_{i=0}^{H_w-1} A_d^i B_d & \cdots & & 0 \\ \vdots & & & \vdots \\ C_z \sum_{i=0}^{H_p-1} A_d^i B_d & \cdots \cdots & & C_z \sum_{i=0}^{H_p-H_u} A_d^i B_d \end{bmatrix}
$$

Then, $Z(k)$ is expressed as Equation (9.4):

$$
Z(k) = \Psi x(k) + \Upsilon u(k-1) + \Theta \Delta U(k) \quad (9.4)
$$

---

3 It is important to become familiar with this sequentially updated representation to understand MPC.

where the objective function of Equation (9.5) is defined as follows[4]:

$$V(k) = \| \mathcal{Z}(k) - \mathcal{T}(k) \|_Q^2 + \| \Delta \mathcal{U}(k) \|_R^2 \tag{9.5}$$

where $\mathcal{T}(k) = \left[r(k + H_w|k) r(k + H_w + 1|k) \cdots r(k + H_p|k) \right]^T$, $Q = diag(Q(H_w)$ $Q(H_w + 1) \cdots Q(H_p))$, and $R = diag(R(0)R(1) \cdots R(H_u - 1))$.

Q and R correspond to the weights, which can be the same for all of the prediction or control horizons, or they can be varied by increasing the weights in the first or second half of the horizon. We also define a deviation function ε expressed by Equation (9.6) as follows:

$$\varepsilon(k) = \mathcal{T}(k) - \Psi x(k) - \Upsilon u(k - 1) \tag{9.6}$$

Then, Equation (9.5) can be rewritten as

$$
\begin{aligned}
V(k) &= \| \mathcal{Z}(k) - \mathcal{T}(k) \|_Q^2 + \| \Delta \mathcal{U}(k) \|_R^2 \\
&\quad \times \| \{ \Psi x(k) + \Upsilon u(k - 1) + \Theta \Delta \mathcal{U}(k) \} - \{ \varepsilon(k) + \Psi x(k) + \Upsilon u(k - 1) \} \|_Q^2 \\
&\quad + \| \Delta \mathcal{U}(k) \|_R^2 \\
&= \| \Theta \Delta \mathcal{U}(k) - \varepsilon(k) \|_Q^2 + \| \Delta \mathcal{U}(k) \|_R^2 \\
&= \left[\Delta \mathcal{U}^T(k) \Theta^T - \varepsilon^T(k) \right] Q \left[\Theta \Delta \mathcal{U}(k) - \varepsilon(k) \right] + \| \Delta \mathcal{U}(k) \|_R^2 \\
&= \varepsilon^T(k) Q \varepsilon(k) - \Delta \mathcal{U}^T(k) \cdot 2\Theta^T Q \varepsilon(k) + \Delta \mathcal{U}^T(k) \left[\Theta^T Q \Theta + R \right] \Delta \mathcal{U}(k) \\
&= Const. - \mathcal{G} \Delta \mathcal{U}(k) + \Delta \mathcal{U}^T(k) \mathcal{H} \Delta \mathcal{U}(k) \tag{9.7}
\end{aligned}
$$

This equation represents the objective function $V(k)$ as a quadratic equation in $\Delta \mathcal{U}(k)$. Since the first term in the last line is independent of $\Delta \mathcal{U}(k)$, the objective function can be rewritten as Equation (9.8) to obtain the optimal $\Delta \mathcal{U}(k)$:

$$V(k) = \Delta \mathcal{U}^T(k) \mathcal{H} \Delta \mathcal{U}(k) - \mathcal{G}^T \Delta \mathcal{U}(k) \tag{9.8}$$

where $\mathcal{H} = \Theta^T Q \Theta + R$, $\mathcal{G} = 2\Theta^T Q \varepsilon(k)$.

A typical model predictive control system, subject to the constraints, finds the change in the control input $\Delta \mathcal{U}(k)$ that minimizes the **objective function** $V(k)$ and then calculates its first component $\Delta \hat{u}(k|k)$, while using $\Delta u(k)$ as $\Delta u(k)$ is repeated for each control cycle. For optimization calculations, it is necessary to solve the **Quadratic Programming Problem**, etc. Here, the fmincon function in MATLAB is used for the optimization calculation.[5]

9.2 Constraint Descriptions

In optimal programming, it is necessary to find the optimal value x that minimizes (or maximizes) the objective function. The constraint equation is expressed as a function of x in this case. Similarly, in designing an MPC system, constraints

---

4 The above refers to $\| x \|_Q^2 = x^T Q x$.

5 Readers may like to know any example of an optimization calculation without the fmincon function. However, we omit it in this book because we do not want the complexity of the optimization calculation to hinder the understanding of the MPC algorithm.

such as the driving forces or torques of a multi-degree-of-freedom mechatronic system, the magnitude of the control input $u(k)$ expressed in terms of velocity in a kinematic model, and the range of motion of a multi-degree-of-freedom mechatronics system must be expressed as functions of $\Delta \mathcal{U}(k)$. To express these constraints with a function of $\Delta \mathcal{U}(k)$ is not always easy, but we will derive it according to the book [2] as follows.

9.2.1 Treatment of Constraints on the Control Input $\hat{u}(k)$

Consider the predictions when the control input $u(k)$ satisfies $-a \leq u(k) \leq a$, $-\hat{u}(k) \leq a$, and $\hat{u}(k) \leq a$, and interpreting $-\frac{1}{a}\hat{u}(k) \leq 1$ and $\frac{1}{a}\hat{u}(k) \leq 1$, we obtain Equation (9.9) for the predicted control input $\hat{u}(k|k) \sim \hat{u}(k+H_u-1|k)$:

$$
\begin{bmatrix}
\frac{-1}{a} & 0 & 0 & \cdots \\
0 & \frac{-1}{a} & 0 & \cdots \\
\vdots & \cdots & \ddots & 0 \\
0 & \cdots & \cdots & \frac{-1}{a} \\
\frac{1}{a} & 0 & 0 & \cdots \\
0 & \frac{1}{a} & 0 & \cdots \\
\vdots & \cdots & \ddots & 0 \\
0 & \cdots & \cdots & \frac{1}{a}
\end{bmatrix}
\begin{bmatrix}
\hat{u}(k|k) \\
\hat{u}(k+1|k) \\
\vdots \\
\hat{u}(k+H_u-1)
\end{bmatrix}
\leq
\begin{bmatrix}
1 \\
1 \\
\vdots \\
1
\end{bmatrix}
\tag{9.9}
$$

where $\hat{u}(k|k) = \Delta\hat{u}(k|k) + u(k-1)\hat{u}(k+1|k) = \Delta\hat{u}(k+1|k) + \hat{u}(k|k) = \Delta\hat{u}(k+1|k) + \{\Delta\hat{u}(k|k) + u(k-1)\}, \ldots$.

By rewriting Equation (9.9) as $\Delta\hat{u}(t|k)$, we obtain a linear inequality constraint, Equation (9.10):

$$
\mathcal{F}\Delta\mathcal{U}(k) \leq -\mathcal{F}_1 u(k-1) + f
\tag{9.10}
$$

where $\mathcal{F} = \begin{bmatrix}
\frac{-1}{a} & 0 & \cdots & 0 \\
\frac{-1}{a} & \frac{-1}{a} & \cdots & 0 \\
\vdots & \cdots & \ddots & 0 \\
\frac{-1}{a} & \frac{-1}{a} & \cdots & \frac{-1}{a} \\
\frac{1}{a} & 0 & \cdots & 0 \\
\frac{1}{a} & \frac{1}{a} & \cdots & 0 \\
\vdots & \cdots & \ddots & 0 \\
\frac{1}{a} & \frac{1}{a} & \cdots & \frac{1}{a}
\end{bmatrix}$, $\Delta\mathcal{U}(k) = \begin{bmatrix}
\Delta\hat{u}(k|k) \\
\Delta\hat{u}(k+1|k) \\
\vdots \\
\Delta\hat{u}(k+H_u-1)
\end{bmatrix}$,

$\mathcal{F}_1 = \begin{bmatrix} \frac{-1}{a} & \frac{-1}{a} & \cdots & \frac{-1}{a} & \frac{1}{a} & \frac{1}{a} & \cdots & \frac{1}{a} \end{bmatrix}^T$, and $f = \begin{bmatrix} 1 & 1 & \cdots & 1 \end{bmatrix}^T$.

9.2.2 Constraints on the Control Variable $\hat{z}(k)$

In the expression for the constraint using $\mathcal{Z}(k)$, the control variable prediction $\hat{z}(k)$ also needs to be rewritten as a constraint on $\Delta \mathcal{U}(k)$. Let $z(k)$ be a scalar value; consider the predicted value when the constraint $-b \leq z(k) \leq c$ is satisfied. Based on $-\hat{z}(k) \leq b$, we interpret $-\frac{1}{b}\hat{z}(k) \leq 1$ and $\frac{1}{c}\hat{z}(k) \leq 1$ and obtain Equation (9.11) with respect to the control variable prediction $\hat{z}(k + H_w|k) \sim \hat{z}(k + H_p|k)$:

$$\Gamma \cdot \mathcal{Z}(k) \leq g \tag{9.11}$$

where $\Gamma = \begin{bmatrix} \frac{-1}{b} & 0 & 0 & \cdots \\ 0 & \frac{-1}{b} & 0 & \cdots \\ \vdots & \cdots & \ddots & 0 \\ 0 & \cdots & \cdots & \frac{-1}{b} \\ \frac{1}{c} & 0 & 0 & \cdots \\ 0 & \frac{1}{c} & 0 & \cdots \\ \vdots & \cdots & \ddots & 0 \\ 0 & \cdots & \cdots & \frac{1}{c} \end{bmatrix}$, and $g = \begin{bmatrix} 1 \\ 1 \\ \vdots \\ 1 \end{bmatrix}$.

Next, using $\mathcal{Z}(k) = \Psi x(k) + \Upsilon u(k-1) + \Theta \Delta \mathcal{U}(k)$ in Equation (9.4), with $\Gamma \cdot \{\Psi x(k) + \Upsilon u(k-1) + \Theta \Delta \mathcal{U}(k)\} \leq g$, we obtain Equation (9.12):

$$\Gamma \Theta \Delta \mathcal{U}(k) \leq -\Gamma \{\Psi x(k) + \Upsilon u(k-1)\} + g \tag{9.12}$$

9.2.3 Constraints on $\Delta \hat{u}(k)$ Change in the Control Input

Consider the constraint on the amount of change in the control input as $-d \leq \Delta u(k) \leq d$ and create Equation (9.13)[6]:

$$\mathcal{E} \Delta \mathcal{U}(k) \leq w \tag{9.13}$$

where $\mathcal{E} = \begin{bmatrix} \frac{-1}{d} & 0 & \cdots & 0 \\ \frac{-1}{d} & \frac{-1}{d} & \cdots & 0 \\ \vdots & \cdots & \ddots & 0 \\ \frac{-1}{d} & \frac{-1}{d} & \cdots & \frac{-1}{d} \\ \frac{1}{d} & 0 & \cdots & 0 \\ \frac{1}{d} & \frac{1}{a} & \cdots & 0 \\ \vdots & \cdots & \ddots & 0 \\ \frac{1}{d} & \frac{1}{d} & \cdots & \frac{1}{d} \end{bmatrix}$, $\Delta \mathcal{U}(k) = \begin{bmatrix} \Delta \hat{u}(k|k) \\ \Delta \hat{u}(k+1|k) \\ \vdots \\ \Delta \hat{u}(k+H_u-1) \end{bmatrix}$, and $w = \begin{bmatrix} 1 & 1 & \cdots & 1 \end{bmatrix}^T$.

---

6 The character d in this section does not mean a disturbance but the maximum and minimum values using characters $a, b, c, \ldots$.

9.2.4 Constraints on the Control Inputs and Quantities

Combining Equations (9.10), (9.12), and (9.13), we obtain Equation (9.14) as linear inequality constraints:

$$
\begin{bmatrix} \mathcal{F} \\ \Gamma\Theta \\ \mathcal{E} \end{bmatrix} \Delta \mathcal{U}(k) \leq \begin{bmatrix} -\mathcal{F}_1 u(k-1) + f \\ -\Gamma\{\Psi x(k) + \Upsilon u(k-1)\} + g \\ w \end{bmatrix} \tag{9.14}
$$

9.3 MPC System Design

Under the conditions of Equation (9.14), a change in the control input, $\Delta \mathcal{U}(k)$, that minimizes the **objective function** $V(k)$ of Equation (9.8) can be obtained. The first data of $\Delta U(k)$ is used as $\Delta u(k)$ repeatedly at each control cycle. For this constrained optimization, a quadratic programming problem is used.

Example 9.3.1 On the cart represented by Equations (4.10) and (4.11), assume the constraints on the control input $|u(k)| \leq u_{max} = 20$ N, position $|x_c(k)| \leq x_{c,max} = 0.80$ m, and a constraint on the amount of change in input $|\Delta u(k)| \leq \Delta u_{max} = 15$ N/s. The position control system is designed with the position reference $r(t|k) = 1.0$ m. The mass of the cart is $m = 2$ kg, viscous friction coefficient $c = 0.1$N/(m/s), control period $T = 10$ms, $H_w = 1$, $H_u = 2$, $H_p = 50$, and $Q = (1 \times 10^6) \times I_{(H_p - H_w + 1)}$ and $R = (1 \times 10^3) \times I_{Hu}$. An illustrative program is shown in List 9.1, and the Simulink model is shown in Figure 9.2.

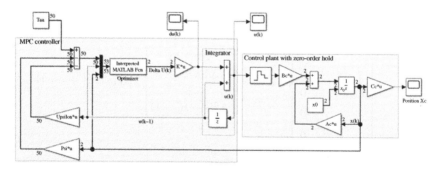

Figure 9.2 Simulink model example of MPC.

List 9.1:   Model predictive control with constraints.

```
1    % T and Hp,Hu,Q,R are quite important adjustment parameters.
2    % Optimization for a range of time T*(Hp−Hw)
3    clear all;clc;
4    %% Definition of variables
5    global xn ul ym delta_U0
6    % xn:Number of states, ul:number of inputs,ym:number of outputs
7    global Hp Hw Hu Q R F F1 f W w Gamma Upsilon Psi Theta g H G_temp options
8    options = optimset('Algorithm','sqp','Display','off');
9    % The above variables should be specified in fmincon if necessary
10
11   %% Physical parameters of the cart and simulation time
12   m=2.0;c=0.1; % Mass of cart and coefficient of viscous friction
13   u_max=20; % Constraints on control inputs − Assume u max<=u<=u max
14   xc_max=0.8; % Constraints on position − assumes xc max<=xc<=xc max
15   du_max=15; % Constraint on input change −assume du max<=du<=du max
16   t_end=5; % Simulation time
17
18   %% State space model
19   Ac=[0,1;0,−c/m];Bc=[0;1/m];Cc=[1,0];
20   [xn,ul]=size(Bc);[ym,~]=size(Cc);Dc=zeros (ym,ul);
21   % xn:Number of states, ul:number of inputs,ym:number of outputs
22   T=0.01;[Ad,Bd,Cd,Dd]=c2dm(Ac,Bc,Cc,Dc,T); % Tmpc:Control period
23
24   %% Determination of MPC adjustment parameters
25   Hw=1;Hu=2;Hp=50;
26   % Hw: window parameters, Hu: control horizon, Hp: predicted horizon
27   Q = eye(ym*(Hp−Hw+1),ym*(Hp−Hw+1))*1e6;
28   R = eye(ul*Hu,ul*Hu)*1e3;
29   K=[ones(1,ul),zeros(1,Hu−ul)]; % Gain to extract only the first value
30
31   %% Design of MPC control parameters
32   %%% Creating Cz %%%
33   Cz=Cd;% Set Cz Set the coefficient matrix as Cz extend.
34   Cz_extend=[];for i=1:Hp−Hw+1 Cz_extend=blkdiag(Cz_extend,Cz);end
35
36   %---- Generating Psi -----
37   Psi_temp=zeros(xn*(Hp−Hw+1),xn);
38   for i=1:Hp−Hw+1;Psi_temp((i−1)*xn+1:i*xn,1:xn)=Ad^(i);end
39   Psi=Cz_extend*Psi_temp;
40
41   %----- Generating Upsilon ----
42   Upsilon_temp = zeros(xn*(Hp−Hw+1),ul);
43   for i = 1:Hp−Hw+1;
```

```
44      for j =1:i
45          for k = 1:xn for l =1:ul up = Ad^(j−1)∗Bd;
46              Upsilon_temp((i−1)∗xn+k,l)=Upsilon_temp((i−1)∗xn+k,l)+up(k,l);
47          end;end;end;end;
48   Upsilon=Cz_extend∗Upsilon_temp;
49
50   %−−−−−−− Generating Theta −−−−−−−
51   Theta_temp=zeros(xn∗(Hp−Hw+1),ul∗Hu); % Theta array
52   for i = 1:Hp−Hw+1
53       for j = 1:Hu
54           if i<j for k =1:xn;
55               for l=1:ul;Theta_temp((i−1)∗xn+k,(j−1)∗ul+l)=0;
56               end;end;
57           else
58               for t=0:(i−j) for k=1:xn for l=1:1:ul;th=Ad^t∗Bd;
59                   Theta_temp((i−1)∗xn+k,(j−1)∗ul+l)=Theta_temp((i−1)∗xn+k,...
60                   (j−1)∗ul+l)+th(k,l);end
61               end;end;end;end;end
62   Theta=Cz_extend∗Theta_temp;
63
64   H=Theta'∗Q∗Theta+R; % Variable used for the objective function
65   G_temp=2∗Theta'∗Q; % Variable used for the objective function
66   % G=G temp∗Epsilon
67
68   %%%%%%%%%%% Generate reference trajectory data Tau %%%%%%%%%%%%%%
69   Tau=[];for j=1:Hp−Hw+1
70       tau_xc=1;Tau=[Tau;tau_xc]; % Position reference (set to 1m)
71   end;
72   disp('Tau size=(number of rows=Hp−Hw+1,number of columns=number of outputs)');
73   disp(size(Tau));
74
75   %% Constraint condition expression:
76   % Generate FGammaTheta, F1, f, etc.(delta_u constraints omitted for simplicity)
77   F=[tril(−1/u_max∗ones(Hu,Hu),0);tril(1/u_max∗ones(Hu,Hu),0)];
78   % tril generates a triangular matrix
79   F1=[−1/u_max∗ones(Hu,1);1/u_max∗ones(Hu,1)];f=ones(2∗Hu,1);
80   Gamma=[−1/xc_max∗eye(Hp−Hw+1);1/xc_max∗eye(Hp−Hw+1)];
81   g=[ones(Hp−Hw+1,1);ones(Hp−Hw+1,1)];
82   W=[tril(−1/du_max∗ones(Hu,Hu),0);tril(1/du_max∗ones(Hu,Hu),0)];w=f;
83
84   %% Initial value of each variable
85   x0=[0;0];u_k_1=0;delta_U0=zeros(Hu,1);%ΔU(0)=0
86
```

```
87   %% Simulation
88   open_system('sim_Figure_9_2_and_9_3_mpc_with_constraints');
89   set_param('sim_Figure_9_2_and_9_3_mpc_with_constraints','WideLines','on');
90   set_param('sim_Figure_9_2_and_9_3_mpc_with_constraints','ShowLineDimensions','on');
91   z=sim('sim_Figure_9_2_and_9_3_mpc_with_constraints');
```

The "interpreted MATLAB function" on the Simulink serves as an optimizer.[7] The integrator is composed of $1/z$ and an adder instead of using the digital integrator provided in the Simulink because it is easy to understand in principle. The choice of the integrator (forward difference, backward difference, and cumulative) is easy to understand because we avoided operations that would differ from the theoretical equation depending on the choice. In List 9.1, $H_p = 50$. Since $T = 0.01$, the control period $T = 0.01$, and it hits $0.01 \times 50 = 0.5$ seconds of the predicted time, almost equal to the rise time of Figure 9.3b. This is important from a practical viewpoint. The control system will diverge if the H_p value is made small unnecessarily to reduce the memory capacity or to make the control program short and simple. If the control period T is made longer, the H_p value can be reduced, but a precise and agile control system cannot be expected. List 9.2 shows the "interpreted MATLAB function" in Figure 9.2.

List 9.2: Optimizer example of model predictive control.

```
1    function delta_U=func_mpc_for_cart(u)
2    global F F1 f Gamma g W w
3    global Psi Upsilon Theta xn ul delta_U0 Hp Hw Hu Epsilon options
4    Epsilon=u(1:Hp−Hw+1);u_k_1=u(Hp−Hw+1+1);
5    x_k=u(Hp−Hw+xn+1:Hp−Hw+xn+2);
6    FGammaThetaW=[F;Gamma*Theta;W]; % description of the coefficient matrix on the left side of
         the constraint
7    F1ufGammagw=[−F1*u_k_1+f;−Gamma*(Psi*x_k+Upsilon*u_k_1)+g;w];
8    .
9    delta_U=fmincon(@objfun_mpc,delta_U0,FGammaThetaW,F1ufGammagw,[],[],[],[],options);
10
11   function y=objfun_mpc(delta_U) % function to write the objective function
12   global H G_temp Epsilon
13   G=G_temp*Epsilon;
14   y=delta_U'*H*delta_U−G'*delta_U; % Objective function
```

In Figure 9.3a, we see that the control input $u(k)$ is constrained within the specified 20 N, and in (b), the position $x_c(k)$ is constrained to 0.8 m.

---

7 Specify the dimension of the output as $Hu * ul$.

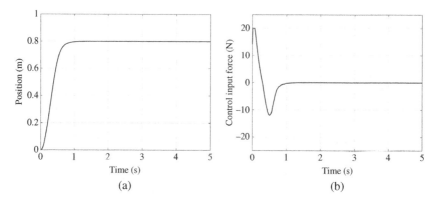

Figure 9.3 An example of MPC simulation. (a) Control input $u(k)$ and (b) position of cart $x_c(k)$.

9.4 Design of Disturbance Observer-Merged MPC System

Considering the configuration to incorporate a disturbance observer, List 9.1 shows the disturbance observer configuration program to be added to List 9.1. Here, assume a one-second width disturbance with 18 N, and it soon changes to a small disturbance with 9 N. This means that the disturbance temporarily exceeds the constraint on the control input. Moreover, we also set the maximum position constraint to 1.2 m. Meanwhile, the maximum value for input change was set to 12 N/s, the control period to 5 ms, $H_u = 10$, and $H_p = 75$. The Simulink example is shown in Figure 9.4, and the response waveforms are shown in Figure 9.5a–c.

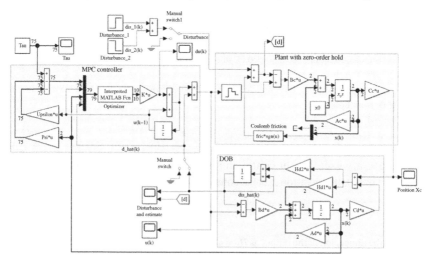

Figure 9.4 Simulink model of disturbance observer-merged MPC.

The characteristic feature of this control system is that the disturbance estimates are taken into the MPC controller, and the control inputs, including the positive feedback values of the disturbances, are used as input constraints.[8]

List 9.3: Model predictive control with disturbance observer (DOB) under constraints 2.

```
1   % T and Hp,Hu,Q,R are quite important adjustment parameters. Optimization for a range of time
        T*(Hp−Hw)
2   %% Definition of variables
3   global xn ul ym delta_U0
4   % xn:Number of states, ul:number of inputs,ym:number of outputs
5   global u_max xc_max du_max
6   global Hp Hw Hu Q R F F1 f Gamma W w
7   global Upsilon Psi Theta g H G_temp options
8   options = optimset('Algorithm','sqp','Display','off');
9   % The above variables should be specified in fmincon if necessary
10
11  %% Physical parameters of the cart and simulation time
12  m=2.0;c=0.1;% Mass of cart and coefficient of viscous friction
13  u_max=20; % Constraints on control inputs − Assume u max<=u<=u max
14  xc_max=1.2; % Constraint on position −assumes xc max<=xc<=xc max
15  du_max=12; % Constraint on input change −assume du max<=du<=du max
16  fric=0.0;t_dis=1.5;dis=18; % Friction and disturbance (Time and magnitude)
17  t_end=5; % Simulation time
18
19  %% State space model
20  Ac=[0,1;0,−c/m];Bc=[0;1/m];Cc=[1,0];
21  [xn,ul]=size(Bc);[ym,~]=size(Cc);Dc=zeros(ym,ul);
22  % xn:Number of states, ul:number of inputs,ym:number of outputs
23  T=0.005;[Ad,Bd,Cd,Dd]=c2dm(Ac,Bc,Cc,Dc,T); % Tmpc:Control period
24
25  %% DOB Design
26  Ad_bar=[Ad,−Bd;zeros(1,2),1];Cd_bar=[Cd,0];
27  pole_dob=[0.2,0.25,0.3];
28  Htemp=place(Ad_bar',Cd_bar',pole_dob);Hd1=Htemp(1:2)';Hd2=Htemp(3);
29
30  %% Determination of MPC adjustment parameters
31  Hw=1;Hu=10;Hp=75;
32  % Hw: window parameters, Hu: control horizon, Hp: predicted hori
33  Q = eye(ym*(Hp−Hw+1),ym*(Hp−Hw+1))*1e7;
34  R = eye(ul*Hu,ul*Hu)*1e1;
35  K=[ones(1,ul),zeros(1,Hu−ul)]; % Gain to extract only the first value
36
```

---

8 In practice, we set u_k_1 to the equation $u\_k\_1 = u(H_p − H_w + 1 + 1) + u(end)$, which is in line 4 of List 9.2, where $u(end)$ corresponds to the disturbance estimate. The need for an MPC design that considers the effects of disturbances can be also found in the literature such as [5].

```
37   %% Design of MPC control parameters
38   %%% Creating Cz %%%
39   Cz=Cd; % Set Cz Set the coefficient matrix as Cz extend
40   Cz_extend=[];for i=1:Hp−Hw+1 Cz_extend=blkdiag(Cz_extend,Cz);end
41
42   %−−−− Generating Psi −−−−−
43   Psi_temp=zeros(xn*(Hp−Hw+1),xn);
44   for i=1:Hp−Hw+1;Psi_temp((i−1)*xn+1:i*xn,1:xn)=Ad^(i);end
45   Psi=Cz_extend*Psi_temp;
46
47   %−−−−− Generating Upsilon −−−−
48   Upsilon_temp = zeros(xn*(Hp−Hw+1),ul);
49   for i = 1:Hp−Hw+1;
50       for j =1:i
51           for k = 1:xn for l =1:ul up = Ad^(j−1)*Bd;
52                   Upsilon_temp((i−1)*xn+k,l)=Upsilon_temp((i−1)*xn+k,l)+up(k,l);
53               end;end;end;end;
54   Upsilon=Cz_extend*Upsilon_temp;
55
56   %−−−−−−− Generating Theta −−−−−−−
57   Theta_temp=zeros(xn*(Hp−Hw+1),ul*Hu); % Theta array
58   for i = 1:Hp−Hw+1
59       for j = 1:Hu
60           if i<j for k =1:xn;
61                   for l=1:ul;Theta_temp((i−1)*xn+k,(j−1)*ul+l)=0;
62                   end;end;
63           else
64               for t=0:(i−j) for k=1:xn for l=1:1:ul;th=Ad^t*Bd;
65                       Theta_temp((i−1)*xn+k,(j−1)*ul+l)=Theta_temp((i−1)*xn+k,...
66                       (j−1)*ul+l)+th(k,l);end
67                   end;end;end;end;end
68   Theta=Cz_extend*Theta_temp;
69
70   H=Theta'*Q*Theta+R; % Variable used for the objective function
71   G_temp=2*Theta'*Q; % Variable used for the objective function
72   % G=G temp*Epsilon
73
74   %%%%%%%%%%% Generate reference trajectory data Tau %%%%%%%%%%%%%%
75   Tau=[];for j=1:Hp−Hw+1
76       tau_xc=1;Tau=[Tau;tau_xc];%;tau_vc];% Position reference (set to 1m)
77   end;
78   disp('Tau size=(number of rows=Hp−Hw+1,number of columns=number of outputs)');
79   disp(size(Tau));
80
```

```
81   %% Constraint condition expression:
82   % Generate FGammaTheta, F1, f, etc.(delta_u constraints omitted for simplicity)
83   F=[tril(-1/u_max*ones(Hu,Hu),0);tril(1/u_max*ones(Hu,Hu),0)];
84   % tril generates a triangular matrix
85   F1=[-1/u_max*ones(Hu,1);1/u_max*ones(Hu,1)];f=ones(2*Hu,1);
86   Gamma=[-1/xc_max*eye(Hp-Hw+1);1/xc_max*eye(Hp-Hw+1)];
87   g=[ones(Hp-Hw+1,1);ones(Hp-Hw+1,1)];
88   W=[tril(-1/du_max*ones(Hu,Hu),0);tril(1/du_max*ones(Hu,Hu),0)];w=f;
89
90   %% Initial value of each variable
91   x0=[0;0];u_k_1=0;delta_U0=zeros(Hu,1);%ΔU(0)=0
92
93   %% Simulation
94   open_system('sim_Figure_9_4_and_9_5_mpc_with_DOB');
95   set_param('sim_Figure_9_4_and_9_5_mpc_with_DOB','WideLines','on');
96   set_param('sim_Figure_9_4_and_9_5_mpc_with_DOB','ShowLineDimensions','on');
97   z=sim('sim_Figure_9_4_and_9_5_mpc_with_DOB');
```

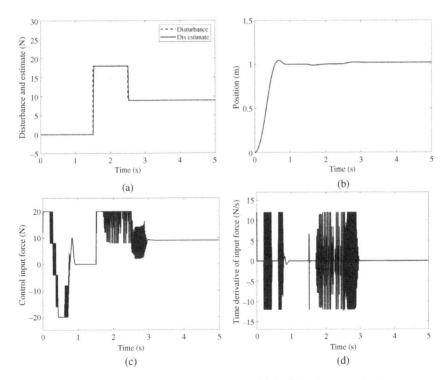

Figure 9.5 Simulation example of DOB-merged MPC. (a) Disturbance and estimate $d(k)$, $\hat{d}(k)$, (b) cart position $x_c(k)$, (c) control input of the cart $u(k)$, and (d) time variation of the control input.

While the disturbance exceeds the constrained value of the control input, the position is disturbed, shown in Figure 9.5b. However, the car's position $x_c(t)$ is almost equal to the reference value after the disturbance decreases, and the control input is within the constraint. Convergence to the reference value is not guaranteed since this example is not a servo system.[9] Figure 9.5c,d also show that the input and input change constraints are obeyed over the entire interval.

Additionally, various design methods are possible, such as the method in [6].

References

1 Jan Brinkhuis, Vladimir Tikhomirov: Optimization: Insights and Applications, Princeton University Press, 2011.

2 Jan Marian Maciejowski: Predictive Control with Constraints, Prentice Hall, 2000.

3 Eduardo F. Camacho, Carlos Bordons Alba: Model Predictive Control, Springer, 2013.

4 Seok-Kyoon Kim, Chang Reung Park, Tae-Woong Yoon, Young IL Lee: Disturbance-observer-based model predictive control for output voltage regulation of three-phase inverter for uninterruptible-power-supply applications, European Journal of Control, Vol. 23, 71–83, 2015.

5 Rié Abe, Toshiyuki Satoh, Naoki Saito, Jun-ya Nagase, Norihiko Saga: Disturbance Observer-Based Model Predictive Control Using Time-Varying Constraints, World Congress on Industrial Control Systems Security (WCICSS-2015), 26–27, 2015.

6 Takashi Ohhira, Akira Shimada, Toshiyuki Murakami: Variable forgetting factor-based adaptive Kalman filter with disturbance estimation considering observation noise reduction, IEEE Access, Vol. 9, 100747–100756, 2021.

---

9 This example is not a servo system, so perfect convergence to the reference value is not guaranteed.

10

Kalman Filter with Disturbance Estimation (KFD)

The **Kalman filter** that includes disturbance estimation may be helpful if the effect of noise is significant and the time constant T in disturbance observer (DOB) must be very long to solve the noise problem. Since Rudolf Emil Kalman introduced the discrete-time Kalman filter in the 1960s, it has been used for satellite orbit estimation in the Apollo program and has been widely used worldwide. Later, the Kalman filter for continuous systems and Kalman filter for nonlinear systems (EKF, UKF, etc.) were also proposed [1–3]. They are used for state estimation of systems with noise. In this chapter, we introduce the Kalman filter using an extended system that incorporates the disturbance as a state variable. We call it "Kalman filter with disturbance estimation" (KFD) [4–6].[1]

10.1 Design of Kalman Filter with Disturbance Estimation

We consider designing a Kalman filter for the state equation (10.1), containing **system noise** $v(k)$, and the output equation (10.2), containing **observation noise** $w(k)$. The system noise $v(k)$ is assumed to be **normality white noise** with mean 0 and variance σ_v^2, and the observed noise $w(k)$ is assumed to be normality white noise with mean 0 and variance σ_w^2. B_v is the coefficient matrix in which the system noise is mixed and is determined by the designer, such as $B_v = I$, when it is assumed that all elements of the state variable $x(k)$ are mixed equally, and $B_v = B_d$, when it is assumed that it is mixed in the control input $u(k)$.

$$x(k + 1) = A_d x(k) + B_d u(k) - B_d d(k) + B_v v(k) \qquad (10.1)$$

---

1 It is not easy to decide whether it is better to deal with the problem by increasing the time constant of the low-pass filter or using the "Kalman filter with disturbance estimation" (KFD). However, we would like to introduce it as one design method.

Disturbance Observer for Advanced Motion Control with MATLAB/Simulink, First Edition. Akira Shimada.
© 2023 The Institute of Electrical and Electronics Engineers, Inc. Published 2023 by John Wiley & Sons, Inc.
Companion website: www.wiley.com/go/disturbanceobserver

$$y(k) = C_d x(k) + w(k) \tag{10.2}$$

Let $\bar{x}(t) = [x(t)^T, d(k)^T]^T$ be the expanded state variable, and the extended system be Equations (10.3) and (10.4):

$$\bar{x}(k + 1) = \bar{A}_d \bar{x}(k) + \bar{B}_d u(k) - \bar{B}_d d(k) + \bar{B}_v v(k) \tag{10.3}$$

$$y(k) = \bar{C}_d \bar{x}(k) + w(k) \tag{10.4}$$

where

$$\bar{A}_d = \begin{bmatrix} A_d & -B_d \\ 0 & I \end{bmatrix}, \bar{B}_d = \begin{bmatrix} B_d \\ 0 \end{bmatrix}, \bar{B}_v = \begin{bmatrix} B_v \\ 0 \end{bmatrix}, \bar{C}_d = [C_d \, | \, 0].$$

Figure 10.1 presents the Kalman filter's block diagram with disturbance estimation. The dashed line at the bottom is the Kalman filter, and the following steps (step 1) and (step 2) are performed in the calculation section. The contents corresponding to step 0 are set as the initial value of z^{-1} connected to $\hat{\bar{x}}(k)$ and the initial value of z^{-1} connected to $P(k)$ in the arithmetic section's output.

The idea of the Kalman filter is to set the expanded state estimation error $\tilde{x}(k) = \bar{x}(k) - \hat{\bar{x}}(k)$ and define the evaluation function $J(k) = E[\tilde{x}^2(k)]$. $J(k)$ corresponds to the **least-squares error estimate**. Let $\bar{x}(k)$ be the $\hat{\bar{x}}(k)$ that minimizes $J(k)$. Finding this $\hat{\bar{x}}(k)$ is called the Kalman filtering problem. Next, we introduce the arithmetic procedure.[2]

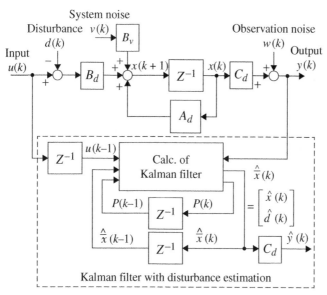

Figure 10.1 Block diagram of the disturbance estimation Kalman filter.

2 For the derivation of the formulas, please refer to specialized books such as [3].

(Step 0) Prepare the following settings first.

(1) Initial state estimates: $\hat{\bar{x}}(0) = E[\bar{x}(0)] = \bar{x}_0$,

(2) Initial value of the covariance matrix: $P(0) = E[(\bar{x}(0) - E[\bar{x}(0)])(\bar{x}(0) - E[\bar{x}(0)])^T] = \Sigma_0$

(Often $P(0) = \gamma \cdot I$ is used.)

(3) Variance of the system noise $v(k)$: σ_v^2 and variance of the observed noise $w(k)$: σ_w^2 For multivariate systems, use the covariance matrix Q instead of σ_v^2 and R instead of σ_w^2.

(Step 1) Prediction step The following calculations are performed.

(1) Prior state estimate: $\hat{\bar{x}}^-(k) = \bar{A}_d \hat{\bar{x}}(k-1) + \bar{B}_d u(k-1)$

(2) Prior error covariance matrix: $P^-(k) = \bar{A}_d P(k-1)\bar{A}_d^T + \sigma_v^2 \bar{B}_v^T \bar{B}_v^T$.

For multivariable systems, $\bar{B}_v Q \bar{B}_v^T$ should be used instead of $\sigma_v^2 \bar{B}_v \bar{B}_v^T$.

(Step 2) Filtering step Perform the following calculations.

(1) Kalman gain: $g(k) = \dfrac{P^-(k)\bar{C}_d^T}{\bar{C}_d P^-(k)\bar{C}_d^T + \sigma_w^2}$

For a multivariate system, $g(k) = P^-(k)\bar{C}_d^T \{\bar{C}_d P^-(k)\bar{C}_d^T + R\}^{-1}C$

(2) State estimate: $\hat{\bar{x}}(k) = \hat{\bar{x}}^-(k) + g(k)\{y(k) - \bar{C}_d \hat{\bar{x}}^-(k)\}$

(3) Posterior error covariance matrix: $P(k) = \{I - g(k)\bar{C}_d\}P^-(k)$

Repeat (Step 1) and (Step 2).

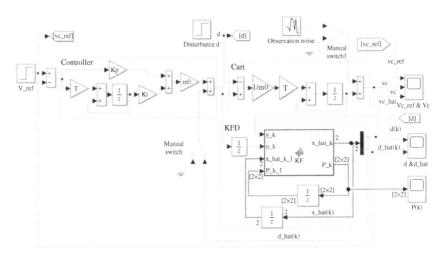

Figure 10.2 Simulink model of Kalman filter with disturbance estimation with velocity observation.

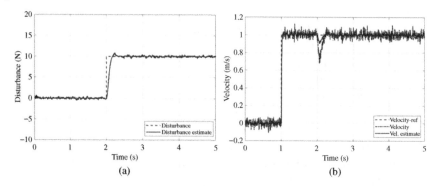

Figure 10.3 Simulation results of disturbance estimation Kalman filter with velocity observation. (a) Disturbance waveforms and (b) velocity waveforms.

Example 10.1.1 (Kalman filter with disturbance estimation with velocity observation.) An example of replacing the observer in a PI[3] velocity control system with a Kalman filter using a digital DOB in the transfer function representation in Section 5.3.3 is presented. An example program is shown in List 10.1, the Simulink model in Figure 10.2, and simulation waveforms in Figure 10.3.

List 10.1: PI velocity control with Kalman filter with disturbance estimation (KFD).

```
1   %% Physical parameters
2   m0=1;d=10; % Mass[kg]) and magnitude of disturbance[N]
3   T=0.01; % Time constant of DOB
4   zeta=7.5;wn=5; % Damping constant, response frequency
5   Kp=2*zeta*wn;Ki=wn^2; % PI controller gains
6
7   %% State space model
8   A=[0,];B=[1/m0];C=[1];D=0;sysc=ss(A,B,C,D);
9   sysd=c2d(sysc,T);[Ad,Bd,Cd,Dd]=ssdata(sysd);X_0=[0];
10
11  %% Extended system Model
12  OA=[Ad,-Bd;0,1];OB=[Bd;0];OC=[Cd,0];OD=0;OBv=eye(2);
13
14  %% Noise Design
15  Qkl=diag([2e-6,2e-2]); % Covariance matrix of hypothetical system noise
16  Rkl=0.001; % Covariance matrix of observation noize
17  dt=0.005; % Time cycle of observation noize
18
19  %% Initial parameters for Kalman Filter Design
```

---

3 See Appendix A.2.2 about PI control.

```
20   gamma=1;P_0=gamma*eye(2); % gamma:initial magnitude of covariance matrix P
21   x_hat_0=[0;0]; % Initial value of state estimate
22
23   %% Simulation
24   tend=5; % Simulation time
25   open_system('sim_Figure_10_2_and_10_3_PI_velocity_control_with_KFD');
26   set_param('sim_Figure_10_2_and_10_3_PI_velocity_control_with_KFD','WideLines','on');
27   set_param('sim_Figure_10_2_and_10_3_PI_velocity_control_with_KFD','ShowLineDimensions','on');
28   y=sim('sim_Figure_10_2_and_10_3_PI_velocity_control_with_KFD');
```

List 10.2:   Example of a MATLAB function in the arithmetic section.

```
1    %------ MATLAB function -----
2    function [x_hat_k,P_k] = Kalman_filter(y_k,u_k,OA,OB,OC,Qkl,Rkl,...
3    x_hat_k_1,P_k_1,OBv)
4
5    %% Priori Estimation
6    x_hat_minus_k=OA*x_hat_k_1+OB*u_k; % Priori State Estimate
7    P_minus_k=OA*P_k_1*OA'+OBv*Qkl*OBv'; % Priori Covariance Matrix
8
9    %% Kalman Gain matrix
10   G_k=P_minus_k*OC'*inv(OC*P_minus_k*OC'+Rkl);% Kalman Gain Matrix
11
12   %% Posteriori Estimation
13   x_hat_k=x_hat_minus_k+G_k*(y_k-OC*x_hat_minus_k); % State Estimates
14   P_k=(eye(2)-G_k*OC)*P_minus_k; % Posterior Covariance Matrix
```

Figure 10.3a shows the disturbance and the estimated waveform. They overlap except for a slight delay. The noise component is smaller than Figure 5.17b-1. Meanwhile, the velocity waveforms and velocity estimates of Figure 10.3b are indistinguishable from those of Figure 5.17b-3, and almost equal speed control characteristics can be achieved.[4]

Example 10.1.2 (Servo system with Kalman filter with disturbance estimation with position observation.)   Let us compare the case of using a Kalman filter with disturbance estimation and a minimal order observer for a digital position servo system. List 10.3 shows an example program for the case using the Kalman filter; a model diagram using Simulink is shown in Figure 10.4; and Figure 10.5a shows

---

4 This design example uses a somewhat unnatural assumption. There is some observed noise, but no system noise is found in Figure 10.2. However, in line 18 of List 10.1, we define a virtual system noise. That is, Q is designed for preparation for a variety of system noises in advance even if no noise now or you may think Q is designed as the tuning parameter, regardless of the real system noise.

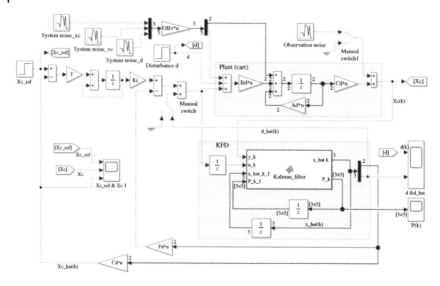

Figure 10.4 Simulink model of a position servo system with a Kalman filter for disturbance estimation in position observation.

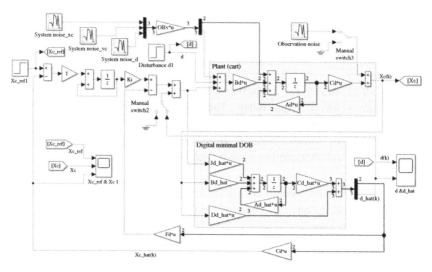

Figure 10.5 Simulink model of position servo system with minimal order digital observer for disturbance estimation in position observation.

the waveforms. Additionally, the contents of the MATLAB function in Figure 10.4 are omitted because the dimension of the unit matrix in the last row of List 10.2 is simply replaced with 3.

List 10.3:   Position servo system with Kalman filter with disturbance estimation.

```
1    %% Physical paremeters
2    m=1;d=10; % Mass[kg] and magnitude of disturbance[N]
3    T=0.01; % Control period[s]
4
5    %% State space model
6    A=[0,1;0,0];B=[0;1/m];C=[1,0];D=0;
7    [Ad,Bd,Cd,Dd]=c2dm(A,B,C,D,T);x0=[0;0];
8
9    %% Creating an Extended System Model for KFD
10   OA=[Ad,−Bd;zeros(1,2),1];OB=[Bd;0];OC=[Cd,0];OD=0;OBv=eye(3);
11
12   %% Noise Design
13   Qx=2e−4;Qv=6e−3;Qd=7e−2;Q=diag([Qx,Qv,Qd]); RR=2e−3;% Real covariances
14   Qklx=Qx;Qklv=Qv;Qkld=Qd;Qkl=diag([Qklx,Qklv,Qkld]); % Covariance matrix for KFD
15   Rkl=RR; % Covariance matrix of ovservation noize for KFD
16   dt=0.005; % Time cycle of observation noize
17
18   %% Initial parameters for Kalman Filter Design
19   gamma=1;P_0=gamma*eye(3); % gamma:initial magnitude of covariance matrix P
20   x_hat_0=[0;0;0]; % Initial value of state estimate
21
22   %% Servo system design
23   Adbar=[Ad,zeros(2,1);−Cd*T,1];Bdbar=[Bd;0];
24   Q=diag([8e2,3e−2,2e4]);R=1;Fd_bar=dlqr(Adbar,Bdbar,Q,R);
25   Fd=Fd_bar(1:2);Ki=−Fd_bar(3);
26
27   %% Simulation
28   open('sim_Figure_10_4_and_10_5_position_servo_system_with_KFD');
29   tend=5;Xc_ref=1; % Simulation time,Position reference
30   set_param('sim_Figure_10_4_and_10_5_position_servo_system_with_KFD','WideLines','on');
31   set_param('sim_Figure_10_4_and_10_5_position_servo_system_with_KFD','ShowLineDimensions','on');
32   y=sim('sim_Figure_10_4_and_10_5_position_servo_system_with_KFD');
```

Next, the program List 10.4 of the servo system with the minimal order digital observer for the disturbance estimation in position observation is shown, but the common parts with List 10.3 are omitted. Figure 10.5 is the corresponding Simulink model.

List 10.4: Position servo system with minimal order digital DOB.

```
1    %% Physical parameters
2    m=1;d=10; % Mass[kg] and magnitude of disturbance[N]
3    T=0.01; % Control period[s]
4
5    %% State space model
6    A=[0,1;0,0];B=[0;1/m];C=[1,0];D=0;
7    [Ad,Bd,Cd,Dd]=c2dm(A,B,C,D,T);x0=[0;0];
8
9    %% Creating an Extended System Model for KFD
10   OA=[Ad,-Bd;zeros(1,2),1];OB=[Bd;0];OC=[Cd,0];OD=0;OBv=eye(3);
11
12   %% Noise Design
13   Qx=2e-4;Qv=6e-3;Qd=7e-2;Q=diag([Qx,Qv,Qd]); RR=2e-3;% Real covariances
14   dt=0.005; % Time cycle of observation noize
15
16   %% Minimal order digital DOB design
17   % Extended state space model fo DOB
18   A_=[A,-B;zeros(1,2),0];B_=[B;0];C_=[C,0];
19   [Ad_,Bd_,Cd_,Dd_]=c2dm(A_,B_,C_,D,T);
20   A11=Ad_(1,1);A12=Ad_(1,2:3);A21=Ad_(2:3,1);A22=Ad_(2:3,2:3);
21   B1=Bd_(1);B2=Bd_(2:3);
22
23   % DOB design
24   pole=[0.90,0.95];Ltemp=place(A22',A12',pole);L=Ltemp';
25   Ad_hat=A22-L*A12,Bd_hat=Ad_hat*L+A21-L*A11
26   Jd_hat=-L*B1+B2,Cd_hat=[zeros(1,2);eye(2)],Dd_hat=[1;L];
27   x_hat_0=[0;0;0]; % Initial value of state estimate
28
29   %% Servo system design
30   Adbar=[Ad,zeros(2,1);-Cd*T,1];Bdbar=[Bd;0];
31   Q=diag([8e2,3e-2,2e4]);R=1;Fd_bar=dlqr(Adbar,Bdbar,Q,R);
32   Fd=Fd_bar(1:2);Ki=-Fd_bar(3);
33
34   %% Simulation
35   tend=5;Xc_ref=1; % Simulation time,Position reference
36   open_system('sim_Figure_10_6_and_10_5b_digital_servo_with_min_DOB');
37   set_param('sim_Figure_10_6_and_10_5b_digital_servo_with_min_DOB','WideLines','on');
38   set_param('sim_Figure_10_6_and_10_5b_digital_servo_with_min_DOB','ShowLineDimensions','on');
39   y=sim('sim_Figure_10_6_and_10_5b_digital_servo_with_min_DOB');
```

Although there is an estimation delay in Figure 10.6a column, the effect of noise is small. When a disturbance is applied, the position deviation can be reduced by adjusting the servo gain, but we dare to use a weaker gain setting. The covariance

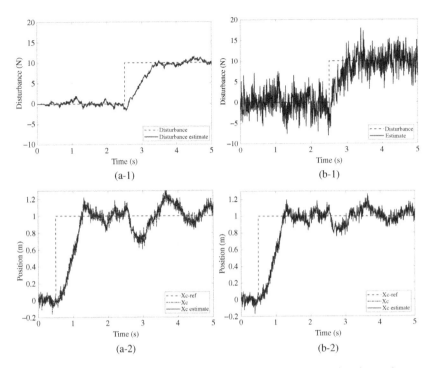

Figure 10.6 Simulation results of Kalman filter with disturbance estimation and minimal order digital DOB in position observation. (a-1) Disturbance of Kalman filter, (a-2) position using Kalman filter, (b-1) disturbance of minimal order digital DOB, and (b-2) position using minimal order digital DOB.

matrix Q_{kl} of the virtual system noise was designed as a tuning parameter, but it tends to prevent fast estimation of the disturbance. The (b) columns are the waveforms of Figure 10.5 with the minimal order digital DOB. Depending on the pole specification, even faster disturbance estimation is possible, but the noise component becomes larger.

Next, let us design a position servo system using a DOB with velocity information. We calculate the velocity from the position information using a pseudo-derivative. Then, we compare the servo system using the real velocity information and the servo system using the above-calculated velocity.

Example 10.1.3 (Position servo system with disturbance observer using velocity calculation.) Servo system design and noise specifications will be the same as in the previous example. An example program is shown in List 10.5, and the model diagrams and waveforms using Simulink are shown in Figures 10.7 and 10.8.

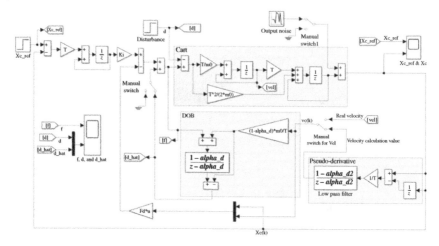

Figure 10.7 Simulink model of a position servo system with a disturbance observer using velocity calculation.

List 10.5: Digital position control with DOB and velocity calculation value.

```
1   %% Physical parameters
2   m0=1;d=10; % Mass[kg] and magnitude of disturbance[N]
3   T=0.01; a=1/T; % Control period[s]
4   alpha_d=exp(-a*T); % Equivalent pole of digital DOB
5   a2=a*5;alpha_d2=exp(-a2*T); % Pole for velocity calculation
6
7   %% State space model
8   A=[0,1;0,0];B=[0;1/m0];C=[1,0];D=0;sysc=ss(A,B,C,D);
9   sysd=c2d(sysc,T);[Ad,Bd,Cd,Dd]=ssdata(sysd);%X_0=[0;0];
10
11  %% Noise Design
12  Rkl=0.001/2; % Covariance matrix of observation noize
13  dt=0.005; % Time cycle of observation noize
14
15  %% Servo system design
16  Adbar=[Ad,zeros(2,1);-Cd*T,1];Bdbar=[Bd;0];
17  Q=diag([8e2,3e-2,2e4]);R=1;Fd_bar=dlqr(Adbar,Bdbar,Q,R);
18  Fd=Fd_bar(1:2);Ki=-Fd_bar(3);
19
20  %% Simulation
21  tend=5;Xc_ref=1; % Simulation time,Position reference
22  open_system('sim_Figure_10_7_digital_pos_con_with_vel_based_DOB');
23  y=sim('sim_Figure_10_7_digital_pos_con_with_vel_based_DOB');
```

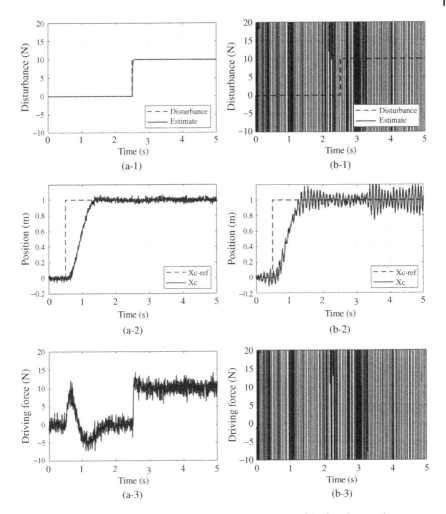

Figure 10.8 Simulation results of a position servo system with disturbance observer using velocity calculation. (a-1) Disturbance estimate using velocity information, (a-2) position using velocity information, (a-3) driving force using velocity information, (b-1) disturbance estimate using calculated velocity values, (b-2) position using calculated velocity values, and (b-3) driving force using calculated velocity.

Figure 10.8a column corresponds that Manual Switch2 in Figure 10.7 sets to the above side. Despite the presence of observation noise, we can see that fast disturbance estimation is achieved in Figure 10.8a-1. The position waveform in Figure 10.8a-2 and the driving force waveform in (a-3) are well controlled.

Meanwhile, the (b) column corresponds that Manual Switch2 in Figure 10.7 sets to the bottom side, and the velocity calculated using pseudo-differentiation is used. The disturbance estimate in the Figure 10.8b-1 figure is buried in noise. The position waveform in Figure 10.8b-2 remains disturbed, and the driving force waveform in (b-3) is also buried in the noise.

Many position control systems use a **rotary encoder** or potentiometer as a position sensor (or angle sensor) and often do not have a velocity sensor. The actual noise arising from rotary encoders is the **quantization error** that occurs for a finite resolution. In this simulation example, we consider a case where the effect of noise is significant, so it may not be such a problem in practice. If you still have a problem, you can replace the sensor with one that has less observation noise, use a high-performance velocity calculation method, or try the proposed Kalman filter or DOB with position observation.[5]

10.2 Design of Stationary Kalman Filter with Disturbance Estimation (SKFD)

A typical Kalman filter seeks an optimal estimate at each control cycle, and the Kalman gain value also changes, but once it converges, the gain value is almost constant. It should be less computationally expensive to use a **stationary Kalman filter** with a constant Kalman gain in such a case [7]. The prior error covariance matrix equation (10.5) given in the design procedure in Section 10.1 can be expressed by substituting the one-sample-lagged posterior error covariance matrix equation (10.6) and the Kalman gain substituting equation (10.7) to obtain Equation (10.8):

$$P^-(k) = \bar{A}_d P(k-1) \bar{A}_d^T + \sigma_v^2 \overline{B}_v \overline{B}_v^T \tag{10.5}$$

$$P(k-1) = \{I - g(k-1)\overline{C}_d\} P^-(k-1) \tag{10.6}$$

$$g(k-1) = \frac{P^-(k-1)\overline{C}_d^T}{\overline{C}_d P^-(k-1)\overline{C}_d^T + \sigma_w^2} \tag{10.7}$$

$$P^-(k) = \bar{A}_d P^-(k-1) \bar{A}_d^T - \bar{A}_d \frac{P^-(k-1)\overline{C}_d^T \overline{C}_d}{\overline{C}_d P^-(k-1)\overline{C}_d^T + \sigma_w^2} P^-(k-1)\bar{A}_d^T$$
$$+ \sigma_v^2 B_v B_v \tag{10.8}$$

For a multivariable system, using $g(k-1) = P^-(k-1)\overline{C}_d^T \{\overline{C}_d P^-(k-1)\overline{C}_d^T + R\}^{-1}$ yields Equation (10.9):

$$P^-(k) = \bar{A}_d P^-(k-1) \bar{A}_d^T$$
$$- \bar{A}_d \{P^-(k-1)\overline{C}_d^T \overline{C}_d\} \{\overline{C}_d P^-(k-1)\overline{C}_d^T + R\}^{-1} P^-(k-1)\bar{A}_d^T$$
$$+ B_v Q B_v \tag{10.9}$$

---

5 High-performance velocity calculation methods are described in Chapter 12.

Here, if $P = P^-(k) = P^-(k-1)$ holds, Equation (10.10) or (10.11), called the algebraic Riccati equation, is obtained.

$$P = \bar{A}_d P \bar{A}_d^T - \frac{\bar{A}_d P \bar{C}_d^T \bar{C}_d P \bar{A}_d^T}{\bar{C}_d P \bar{C}_d^T + \sigma_w^2} + \sigma_v^2 B_v B_v \qquad (10.10)$$

$$P = \bar{A}_d P \bar{A}_d^T - \bar{A}_d P \bar{C}_d^T (\bar{C}_d P \bar{C}_d^T + R)^{-1} \bar{C}_d P \bar{A}_d^T + B_v Q B_v \qquad (10.11)$$

If these are positive-definite symmetric solutions, the Kalman gain is Equation (10.12), and for multivariable systems, it is Equation (10.13):

$$g = \frac{P \bar{C}_d^T}{\bar{C}_d P \bar{C}_d^T + \sigma_w^2} \qquad (10.12)$$

$$g = P \bar{C}_d^T \{\bar{C}_d P \bar{C}_d^T + R\}^{-1} \qquad (10.13)$$

Example 10.2.1 (Servo system with a stationary Kalman filter for disturbance estimation with position observation.) A stationary Kalman filter with disturbance estimation using position observation is designed as a servo system with the cart model. An example program is shown in List 10.6, and a model diagram using rm Simulink is shown in Figure 10.9. All physical parameters are shown in the following program; they are the same throughout this section.

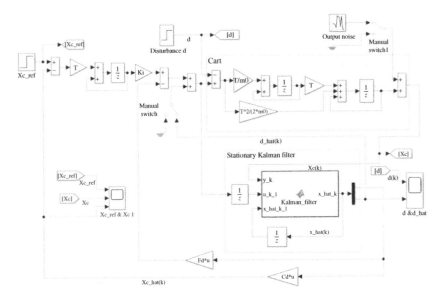

Figure 10.9 Simulink model of a position servo system with disturbance estimation stationary Kalman filter.

List 10.6: Position control system with stationary Kalman filter.

```
1   %% Physical paremater
2   m0=1;d=10; % Mass[kg] and magnitude of disturbance[N]
3   T=0.01; % Control period[s]
4
5   %% State space model
6   A=[0,1;0,0];B=[0;1/m0];C=[1,0];D=0;
7   [Ad,Bd,Cd,Dd]=c2dm(A,B,C,D,T);X_0=[0;0];
8
9   %% Creating an Extended System Model for KFD
10  OA=[Ad,-Bd;zeros(1,2),1];OB=[Bd;0];OC=[Cd,0];OD=0;OBv=eye(3);
11
12  %% Noise Design
13  Qklx=2e2;Qklv=6e-3;Qkld=7e4;Qkl=diag([Qklx,Qklv,Qkld]);
14  % Covariance matrix of hypothetical system noise for KFD
15  Rkl=0.001/2; % Covariance matrix of observation noize
16  dt=0.005; % Time cycle of observation noize
17
18  %% stationary Kalman Filter Design
19  set(gca,'Fontname','Times New Roman','FontSize',14); % Characters setting
20  set(gcf,'color','w'); % Change window into white color
21  %
22  [H_temp,P,e]=dlqr(OA',OC',Qkl,Rkl);G_k=H_temp';
23  ee=P-OA*P*OA'+OA*P*OC'*inv(Rkl+OC*P*OC')*OC*P*OA'-Qkl;
24  x_hat_0=[0;0;0]; % Initial value of state estimate
25
26  %% Servo system design
27  Adbar=[Ad,zeros(2,1);-Cd*T,1];Bdbar=[Bd;0];
28  Q=diag([8e2,3e-2,2e4]);
29  R=1;Fd_bar=dlqr(Adbar,Bdbar,Q,R);
30  Fd=Fd_bar(1:2);Ki=-Fd_bar(3);
31
32  %% Simulation
33  tend=5;Xc_ref=1; % Simulation time,Position reference
34  open_system('sim_Figure_10_9_and_10_10_pos_con_with_Stationary_KF');
35  set_param('sim_Figure_10_9_and_10_10_pos_con_with_Stationary_KF','WideLines','on');
36  set_param('sim_Figure_10_9_and_10_10_pos_con_with_Stationary_KF','ShowLineDimensions','on');
37  y=sim('sim_Figure_10_9_and_10_10_pos_con_with_Stationary_KF');
```

The program List 10.7 is the code in Kalman_filter (MATLAB function) of Figure 10.9.

List 10.7: MATLAB function of steady Kalman filter.

```
1   function x_hat_k = Kalman_filter(y_k,u_k_1,G_k,OC,x_hat_k_1, OA, OB)
2   x_hat_minus_k=OA*x_hat_k_1+OB*u_k_1;
3   x_hat_k=x_hat_minus_k+G_k*(y_k-OC*x_hat_minus_k); % State estimate
```

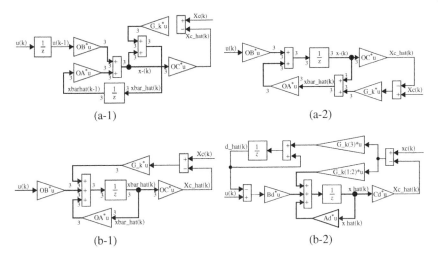

Figure 10.10 Comparison of the structures of disturbance estimation steady-state Kalman filter and digital DOB in position observation. (a-1) Block diagram of a steady Kalman filter, (a-2) figure organized by $1/z$ together, (b-1) digital DOB with LQ design, and (b-2) separation of state variables and disturbances.

The simulation results were equivalent to the column of Figure 10.6a and are therefore omitted. We can replace the content of the stationary Kalman filter enclosed by the dashed line in Figure 10.9 with that of the stationary Kalman filter in Figure 10.10a-1. If the two delay elements $1/z$ are equivalently collected and organized in one place, (a-1) becomes (a-2). This is a simple representation of the disturbance estimation stationary Kalman filter.

Meanwhile, Figure 10.8(b) column represents the digital DOB. When the extended state variable $\bar{\hat{x}} = [\hat{x}, \hat{d}]^T$ is separately represented, it becomes (b-2) of the form described in Section 10.1. The two are almost equivalent, using the same Q and R values as the Kalman filter to determine the observer gain G_k. However, a close comparison of (a-2) and (b-1) shows that the input of the expanded C matrix (OC) is different. It can be observed that the output of the Kalman gain or observer gain is connected to a different destination before or after the delay element $1/z$.[6] There appears to be no particular preference, and the choice is left to the designer.

10.3 Design of Extended Kalman Filter with Disturbance Estimation (EKFD)

Known state estimation methods for nonlinear systems include **extended Kalman filter** (EKF) and unscented Kalman filter (UKF). For the state variables

---

6 G_k is used in this example for comparison, although this book usually expresses an observer gain by H.

$x(k) \in R^n$, the control inputs $u(k) \in R^m$, the observed outputs $y[k] \in R^l$, the mean value of the system noise $v[k]$ is 0, and its variance σ_v^2. Assume that the mean of the observation noise $w[k]$ is 0, and its variance is σ_w^2. The state equation and output equations become Equations (10.14) and (10.15):

$$x(k + 1) = f(x(k)) + B_d u(k) - B_d d(k) + B_v v(k) \tag{10.14}$$

$$y(k) = h(x(k)) + w(k) \tag{10.15}$$

Let $\bar{x}(t) = [x(t)^T, d(k)^T]^T$ denote the extended state variable, and the extended system be Equations (10.16) and (10.17):

$$\bar{x}(k + 1) = \bar{f}(\bar{x}(k)) + \bar{B}_d u(k) - \bar{B}_d d(k) + \bar{B}_v v(k) \tag{10.16}$$

$$y(k) = \bar{h}(\bar{x}(k)) + w(k) \tag{10.17}$$

The basic EKF design process is as follows:

(Step 1) Prior estimation Prior state estimate $\hat{x}^-(k) = f(\hat{x}(k-1)) + B_d u(k-1)$.

(Step 2) Linear approximation and prior error covariance matrix $A_d(k-1) = \frac{\partial f(x)}{\partial x}|_{x=\hat{x}(k-1)}$, as $C_d^T(k) = \frac{\partial h(x)}{\partial x}|_{x=\hat{x}^-(k)}$ Find the prior error covariance matrix.
$P^-(k) = A_d(k-1)P(k-1)A_d^T(k-1) + \sigma_v^2 B_d B_d^T$.

(Step 3) Kalman gain $g(k) = P^-(k)C_d^T(k)\{C_d(k)P^-(k)C_d^T(k) + \sigma_w^2\}^{-1}$.

(Step 4) Posterior estimation State estimate $\hat{x}(k) = \hat{x}^-(k) + g(k)\{y(k) - h(\hat{x}^-(k))\}$. Posterior error covariance matrix $P(k) = \{I - g(k)C_d(k)\}P^-(k)$.

Example 10.3.1 As a simple example, consider the one-link manipulator affected by the gravity field shown in Figure 10.11. Denoting the manipulator tip mass m, arm length r, rotation angle $\theta(t)$, rotation speed $\omega(t) = \dot{\theta}(t)$, gravitational acceleration g, drive torque $\tau(t)$, and disturbance $\tau_d(t)$, the equation of motion is

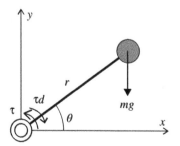

Figure 10.11 Image of one-link manipulator.

given by Equation (10.18). Then, assume that the plant receives the noises $v(t)$ and $w(t)$. Here, $v(t)$ is the system noise with the mean 0, variance σ_v^2, and $w(t)$ is the observation noise with the mean 0, variance σ_w^2:

$$mr^2\dot{\omega}(t) + mgr\cos\theta(t) = \tau(t) - \tau_d(t) \tag{10.18}$$

Let the state variable be $x(t) = [\theta(t), \omega(t)]^T$, the control input $u(t) = \tau(t)$, the observed output $y(t) = \theta(t)$, and noises are added, Equations (10.19) and (10.20) become the state and output equations:

$$\begin{bmatrix} \dot{\theta}(t) \\ \dot{\omega}(t) \end{bmatrix} = \begin{bmatrix} \omega(t) \\ -\frac{g}{r}\cos\theta(t) \end{bmatrix} + \begin{bmatrix} 0 \\ \frac{1}{mr^2} \end{bmatrix} \tau(t)$$

$$- \begin{bmatrix} 0 \\ \frac{1}{mr^2} \end{bmatrix} \tau_d(t) + \begin{bmatrix} 0 \\ \frac{1}{mr^2} \end{bmatrix} v(t) \tag{10.19}$$

$$y(t) = \theta = \begin{bmatrix} 1 & 0 \end{bmatrix} \begin{bmatrix} \theta(t) \\ \omega(t) \end{bmatrix} + w(t) \tag{10.20}$$

Since the digital system's EKF cannot be designed directly from this continuous model, we should derive the digital model as follows. Let the time derivative $\dot{\theta}(t)$ be $\{\theta(k+1) - \theta(k)\}/T$ and $\dot{\omega}(t)$ be $\{\omega(k+1) - \omega(k)\}/T$, and rearrange them as follows.

$$\begin{bmatrix} \theta(k+1) \\ \omega(k+1) \end{bmatrix} = \begin{bmatrix} \theta(k) + T\omega(k) \\ \omega(k) - \frac{Tg}{r}\cos\theta(k) \end{bmatrix} + \begin{bmatrix} 0 \\ \frac{T}{mr^2} \end{bmatrix} \tau(k)$$

$$- \begin{bmatrix} 0 \\ \frac{T}{mr^2} \end{bmatrix} \tau_d(k) + \begin{bmatrix} 0 \\ \frac{T}{mr^2} \end{bmatrix} v(k) \tag{10.21}$$

$$\theta(k) = \begin{bmatrix} 1 & 0 \end{bmatrix} \begin{bmatrix} \theta(k) \\ \omega(k) \end{bmatrix} + w(k) \tag{10.22}$$

Since the first term on the right-hand side of Equation (10.21) corresponds to $f(x(k))$, and the coefficient matrix of the first term on the right-hand side of Equation (10.22) corresponds to $h(x(k))$, the A_d and C_d matrices in (Step 2) are expressed as Equations (10.23) and (10.24).

$$A_d(k-1) = \frac{\partial f(x)}{\partial x}\Big|_{x=\hat{x}(k-1)} = \begin{bmatrix} \frac{\partial f(x)}{\partial \theta} & \frac{\partial f(x)}{\partial \omega} \end{bmatrix}_{x=\hat{x}(k-1)}$$

$$= \begin{bmatrix} 1 & T \\ \frac{Tg}{r}\sin\theta(k) & 1 \end{bmatrix} \tag{10.23}$$

$$C_d = \begin{bmatrix} 1 & 0 \end{bmatrix} \tag{10.24}$$

In this example, the EKF ignores the disturbance, leading to estimation errors for the disturbance. Therefore, we design an EKF that simultaneously estimates the disturbance. Let the state variable $\bar{x}(t) = [\theta(t), \omega(t), \tau_d(k)]^T$, the control input $u(t) = \tau(t)$, and the observed output $y(t) = \theta(t)$, the state and output equations of the expanded system can be expressed as Equations (10.25) and (10.26). However, $\dot{\tau}_d(t) = 0$, whereas $\tau_d(k+1) = \tau_d(k)$ in the digital system for the step disturbance:

$$
\begin{bmatrix} \dot{\theta}(t) \\ \dot{\omega}(t) \\ \dot{\tau}_d(t) \end{bmatrix} = \begin{bmatrix} \omega(t) \\ -\frac{g}{r}\cos\theta(t) - \frac{1}{mr^2}\tau_d(t) \\ 0 \end{bmatrix} + \begin{bmatrix} 0 \\ \frac{1}{mr^2} \\ 0 \end{bmatrix} \tau(t)
$$

$$
+ \begin{bmatrix} 0 \\ \frac{1}{mr^2} \\ 0 \end{bmatrix} v(t) \tag{10.25}
$$

$$
y(t) = \theta = \begin{bmatrix} 1 & 0 & 0 \end{bmatrix} \begin{bmatrix} \theta(t) \\ \omega(t) \\ \tau_d(t) \end{bmatrix} + w(t) \tag{10.26}
$$

As in the previous example, the digital model is Equations (10.27) and (10.28):

$$
\begin{bmatrix} \theta(k+1) \\ \omega(k+1) \\ \tau_d(k+1) \end{bmatrix} = \begin{bmatrix} \theta(k) + T\omega(k) \\ \omega(k) - \frac{Tg}{r}\cos\theta(k) - \frac{T}{mr^2}\tau_d(k) \\ \tau_d(k) \end{bmatrix}
$$

$$
+ \begin{bmatrix} 0 \\ \frac{T}{mr^2} \\ 0 \end{bmatrix} \tau(k) + \begin{bmatrix} 0 \\ \frac{T}{mr^2} \\ 0 \end{bmatrix} v(k) \tag{10.27}
$$

$$
y(k) = \theta(k) = \begin{bmatrix} 1 & 0 & 0 \end{bmatrix} \begin{bmatrix} \theta(k) \\ \omega(k) \\ \tau_d(k) \end{bmatrix} + w(k) \tag{10.28}
$$

The corresponding $A_d(k-1)$ and $C_d(k)$ are as follows:

$$
A_d(k-1) = \frac{\partial f(x)}{\partial x}\big|_{x=\hat{x}(k-1)} = \begin{bmatrix} 1 & T & 0 \\ \frac{Tg}{r}\sin\theta(k-1) & 1 & -\frac{T}{mr^2} \\ 0 & 0 & 1 \end{bmatrix}
$$

$$
C_d(k) = \begin{bmatrix} 1 & 0 & 0 \end{bmatrix} \tag{10.29}
$$

Next, an example program for the EKF design is shown in List 10.8, an example model in MATLAB/Simulink is shown in Figure 10.12, and a simulation example is shown in Figure 10.13. In addition, Lists 10.9 and 10.10 are the MATLAB functions in Figure 10.12.

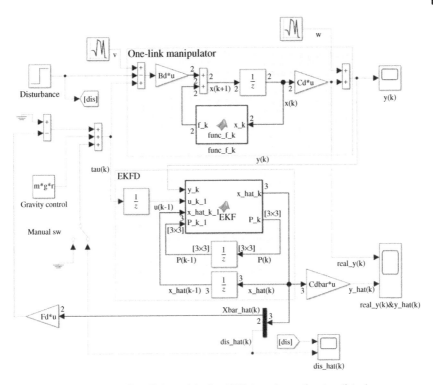

Figure 10.12 Example Simulink model of an EKF that also estimates disturbances.

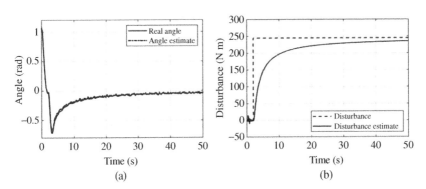

Figure 10.13 Simulation example of control system with disturbance estimation EKF. (a) Position and position estimates and (b) disturbances and disturbance estimates.

List 10.8: EKFD for 1 link manipulator.

```
1    %-----------------------------------------------------------
2    % Xbar(k)=[theta(k);omega(k);
3    % dis(k)=[angle;angular velocity;disturbance];
4    % f(x)=[theta(k)+T*omega(k);
5    % omega(k)-Tg/r*cos(theta(k))-1/(mr^2)*dis(k);
6    % dis(k) ];
7    % Ad=[1 ,T, 0;
8    % Tg/r*sin(theta_k) ,1, -T/(mr^2);
9    % 0 ,0, 1 ];
10   % Bd=[0; T/(m*r^2);0];Cd=[1 0 0];
11   %-----------------------------------------------------------
12
13   %% Physical parameters
14   m=10;r=2.5;g=9.8;c=0.0; % Mass,radius,gravity accel, viscous friction
15   dis=m*r*g*1.0; % Magnitude of disturbance
16   dis_time=2.0; % Start time of disturbance
17   T=0.01;tend=50; % Control period and simulation time
18   sigma_v_2=1e-3;sigma_w_2=1e-3; % Real covariances of system and observation noize
19   Qkf=diag([1e-1,1e1,1e7]);
20   Rkf=1e-6;% Covariances of system and ovservation noize in EKFD
21   theta_0=pi/3;omega_0=0; x_0=[theta_0;omega_0];% Initial values
22
23   %% Preparation of EKFD design
24   Bd=[0;T/(m*r^2)];Cd=[1 0];
25   Bdbar=[Bd;0];Cdbar=[Cd 0];Bv=eye(3);
26   theta_hat_0=0;omega_hat_0=0;dis_hat_0=0;
27   x_hat_0=[theta_hat_0;omega_hat_0;dis_hat_0];
28   P_0=1*eye(3);% Initial value of error covariance matrix
29
30   %% Control system design
31   Q=diag([3e4,2e3]);R=1;
32   Adf=[1 ,T;
33       T*g/r*sin(theta_0),1];
34   Fd=dlqr(Adf,Bd,Q,R);
35
36   %% Simulation
37   open_system('sim_Figure_10_12_and_10_13_EKFD_for_1link_manipulator');
38   set_param('sim_Figure_10_12_and_10_13_EKFD_for_1link_manipulator','WideLines','on');
39   set_param('sim_Figure_10_12_and_10_13_EKFD_for_1link_manipulator',...
40   'ShowLineDimensions','on');
41   z=sim('sim_Figure_10_12_and_10_13_EKFD_for_1link_manipulator');
```

List 10.9: Example of $f(k)$ in one link manipulator.

```
1  function f_k = func_f_k(x_k,T,g,r)
2  %.
3  theta_k=x_k(1);omega_k=x_k(2);
4  f_k=[theta_k+T*omega_k;
5  omega_k–T*g/r*cos(theta_k)];
```

List 10.10: EKF example of a linked manipulator.

```
1
2  function [x_hat_k,P_k] =
3   EKF(y_k,u_k_1,x_hat_k_1,P_k_1,Rkf,Qkf,Bdbar,Cdbar,T,g,r,m)
4  %.
5  theta_k=y_k;theta_k_1=y_k;
6  theta_hat_k_1=x_hat_k_1(1);omega_hat_k_1=x_hat_k_1(2);
7  dis_hat_k_1=x_hat_k_1(3);
8
9  %% prior estimate
10 x_hat_minus_k=[theta_hat_k_1+T*omega_hat_k_1;
11 omega_hat_k_1–T*g/r*cos(theta_hat_k_1)–T/(m*r^2)*dis_hat_k_1;
12 dis_hat_k_1]+Bdbar*u_k_1;
13
14 %% Kalman gain calculation and posterior estimate
15 Ad_k_1=[1,T,0;
16 T*g/r*sin(theta_k_1),1,–T/(m*r^2);
17 0,0,1]; % Derivation of Ad(k) matrix by Taylor expansion
18 Cd_k=Cdbar;
19 %
20 P_minus_k=Ad_k_1*P_k_1*Ad_k_1'+Qkf*Bdbar*Bdbar';
21 %
22 g_k=P_minus_k*Cd_k'/(Cd_k*P_minus_k*Cd_k'+Rkf); % Kalman gain
23 x_hat_k = x_hat_minus_k+g_k*(y_k–Cd_k*x_hat_minus_k); % posterior estimate
24 P_k=(eye(3)–g_k*Cd_k)*P_minus_k; % Update P matrix
```

The slow disturbance estimation can be seen from the output waveform in Figure 10.13(b). Although there is still room for further work, such as increasing the order of the disturbance model, it can be seen that an estimation function that considers the effect of noise has been realized. Since the disturbance estimate in this EKF estimates only the external force and excludes the gravity estimate, the estimate's positive feedback does not result in a gravity compensation. Therefore, the gravity compensation $m * g * r$ is added in the Simulink model as gravity control. The joint angles converge to zero when used in conjunction with the disturbance estimate's positive feedback.

The characteristics of the EKF for disturbance estimation in this section are not good. It delays disturbance estimation, so it can be said that this method is still under research. However, we introduced it as an estimation method for nonlinear systems when noise cannot be ignored, expecting that it will be developed in the future.

References

1 Rudolf Emil Kalman: A new approach to linear filtering and prediction problems, Transactions of ASME-Journal of Basic Engineering, Vol. 82, 35–45, 1960.

2 Simon Julier, Jeffrey Uhlmann, Hugh F. Durrant-Whyter: A new method for the nonlinear transformation of means and covariances in filters and estimators, IEEE Transactions on Automatic Control, Vol. 45, No. 3, 477–482, 2000.

3 Mohinder S. Grewal, Angus P. Andrews: Kalman Filtering: Theory and Practice with MATLAB, Wiley-IEEE Press, 2014.

4 Akira Shimada: Kalman Filter Design with Disturbance Estimation, IEEJ Industrial Electronics Technical Meeting, MEC-16-27, 2016 (In Japanese).

5 Takashi Ohhira, Akira Shimada: Movement control based on model predictive control with disturbance suppression using Kalman filter including disturbance estimation, IEEJ Journal of Industry Applications, Vol. 7, No. 5, 387–395, 2018.

6 Takashi Ohhira, Akira Shimada, Toshiyuki Murakami: Variable forgetting factor-based adaptive Kalman filter with disturbance estimation considering observation noise reduction, IEEE Access, Vol. 9, 100747–100756, 2021.

7 Akira Shimada, Chaisamorn Yongyai: Motion control of inverted pendulum robots using a Kalman filter based disturbance observer, Transactions of SICE JCMSI, Vol. 2, No. 1, 50–55, 2009.

11

Adaptive Disturbance Observer

Consider an **adaptive disturbance observer** that simultaneously estimates the parameters, state variables, and disturbances of a control plant with unknown parameters [1]. This section shows a design process of an adaptive disturbance observer by making use of a Kreisselmeier-type **adaptive observer** [2].

However, since this book does not focus on **adaptive control**, we do not extend the discussion to recent robust adaptive control theory, simple adaptive control, etc. [3]. Additionally, several conditions, including observability, are necessary for designing adaptive observers. Still, this chapter only introduces the basic design flow and design examples based on the assumption of designable plants.

11.1   Structure of an Adaptive Observer

First, let us introduce the general structure of an adaptive observer and understand the path to designing each element of the observer in sequence.

Suppose that a 1-input 1-output n-order control plant is represented by $P(s) = [A, B, C, 0]$. We introduce the way to express this control plant by the **state variable filter** and unknown parameters. The **unknown parameters**, $\hat{\phi}$ value, have been calculated from the difference between the output estimate, $\hat{y}(t)$, and the real output $y(t)$. The state variable estimate $\hat{x}(t)$ is then calculated from state variable filter output $\xi(t)$ and the unknown parameter estimate $\hat{\phi}$. Figure 11.1 shows the structure. However, the adaptive observer design does not directly estimate the physical parameters of the state equation created from the equation of motion because an observable canonical system represents the control plant, and this method estimates the unknown parameters in the canonical system. Additionally, the state variables $z(t)$ of the **observable canonical system** have no physical meaning, so if the observable canonical system was obtained by some

Disturbance Observer for Advanced Motion Control with MATLAB/Simulink, First Edition. Akira Shimada.
© 2023 The Institute of Electrical and Electronics Engineers, Inc. Published 2023 by John Wiley & Sons, Inc.
Companion website: www.wiley.com/go/disturbanceobserver

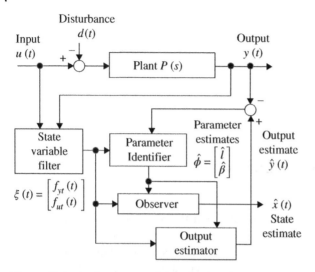

Figure 11.1 Basic structure of the adaptive observer.

coordinate transformation, if necessary, it is necessary to return it to the original state variables $x(t)$.[1]

11.2 Derivation of Observable Canonical System for Adaptive DOB

Referring to Chapter 4, we assume an input disturbance with $\dot{d}(t) = 0$ for a controllable and observable one-input one-output system, $\bar{x}(t) = [x^T(t), d(t)]^T \in R^{n+1}$ and Equation (11.1), consider the expanded system of Equation (11.2). Hereafter, let $\bar{n} = n + 1$:

$$\begin{bmatrix} \dot{x} \\ \dot{d} \end{bmatrix} = \begin{bmatrix} A & -B \\ 0 & 0 \end{bmatrix} \begin{bmatrix} x \\ d \end{bmatrix} + \begin{bmatrix} B \\ 0 \end{bmatrix} u \tag{11.1}$$

$$y = \begin{bmatrix} C & 0 \end{bmatrix} \begin{bmatrix} x \\ d \end{bmatrix} \tag{11.2}$$

---

1 When the observable canonical system state variable estimates $\hat{z}(t)$ are used for control as they are, the output of the state observer of Figure 11.1 should be $\hat{z}(t)$ instead of $\hat{x}(t)$. Moreover, since this document deals with an extended system that includes disturbances as state variables, assume $\hat{\bar{x}}(t) = [\hat{x}(t)^T, \hat{d}(t)]^T$ is the state estimate in Section 11.2 and thereafter.

The coefficient matrix of the first term on the right-hand side of Equation (11.1) is $\bar{A} \in R^{\bar{n} \times \bar{n}}$, and that of the second term is $\bar{B} \in R^{\bar{n} \times 1}$. Let $\bar{C} \in R^{1 \times \bar{n}}$ be the coefficient matrix on the right side of Equation (11.2).

Denote the above equation as $\dot{\bar{x}} = \bar{A}\bar{x} + \bar{B}u, y = \bar{C}\bar{x}$. The transfer function is $\bar{P}(s) = \bar{C}(sI - \bar{A})^{-1}\bar{B}$, which is specifically considered to be in the minimal realization form equivalent to Equation (11.3)[2]:

$$\bar{P}(s) = \frac{\beta_1 s^{\bar{n}-1} + \cdots + \beta_n}{s^{\bar{n}} + \alpha_1 s^{\bar{n}-1} + \cdots + \alpha_{\bar{n}}} \tag{11.3}$$

Next, from the characteristic equation of $\bar{A}$, a matrix W is created using polynomial coefficients and multiplied with the observability matrix U_0 to obtain the observable canonical system.[3]

However, since the layout of the matrices differs from that of the observable canonical system presented in many textbooks, we perform a transformation to adjust the order and obtain Equations (11.4) and (11.5):

$$\dot{\bar{z}}(t) = \tilde{A}\bar{z}(t) + \tilde{B}u(t) = \begin{bmatrix} -\alpha & g^T \\ & L \end{bmatrix} \bar{z}(t) + \tilde{B}u(t) \tag{11.4}$$

$$y(t) = \tilde{C}\,\bar{z}(t) \tag{11.5}$$

g denotes $\bar{n} - 1$ order known vector, and L denotes $\bar{n} - 1 \times \bar{n} - 1$ known matrix[4]:

$$g^T = [1, 0, \ldots, 0], \qquad L = \begin{bmatrix} 0 & I_{\bar{n}-2} \\ \hline 0 & 0 \cdots 0 \end{bmatrix} \tag{11.6}$$

The parameters of Equations (11.4) and (11.5) are $\alpha = [\alpha_1, \ldots, \alpha_{\bar{n}}]^T$, and there are $2\bar{n}$ of them with $\tilde{B} = \beta = [\beta_1, \ldots, \beta_{\bar{n}}]^T$. These are unknown constants, and we aim to identify them adaptively.

11.3 Creating State Variable Filter

Considering that all $2\bar{n}$ coefficients of the transfer function are unknown, it is necessary to devise a way to estimate them. Then, we need to design a state

---

2 Controllable and observable state and output equations can be created from the transfer function. The pair of the equations and the output equation is of the same dimension as the transfer function and is called the minimal realization.

3 In general, coordinate transformation is required to obtain an observable normal form, so create a transformation matrix $S_1 = WU_0$ using the coefficient matrix W and the observable matrix U_0, and get $\bar{\bar{x}}(t) = S_1^{-1}z(t)$. Get $\dot{\bar{\bar{x}}}(t) = S_1^{-1}\dot{z}(t) = AS_1^{-1}z(t) + Bu(t)$. Furthermore, multiply both sides by S_1 from the left, $\dot{z}(t) = \bar{A}z(t) + \bar{B}u(t)$ and get $y(t) = CS_1^{-1}z(t) = \bar{C}z(t)$. Here, $\bar{A} = S_1 A S_1^{-1}$ and $\bar{B} = S_1 B$. Next, for the convenience of design, create a transformation matrix S_2 and set $z(t) = S_2^{-1}\bar{z}(t)$ for the same transformation. Then, Equations (11.4) and (11.5) are obtained. $\tilde{A} = (S_2 S_1)A(S_2 S_1)^{-1}$, $\tilde{B} = (S_2 S_1)B$, and $\tilde{C} = C(S_2 S_1)^{-1}$.

4 In general, H is called the $n - 1$th known vector, and L is defined as the $(n - 1) \times (n - 1)$ matrix, where $\bar{n} - 1 = (n + 1) - 1 = n$ in this section.

variable filter in the form of **nonminimal realization** instead of a **state variable filter**. Let us follow the literature [4] and expand against the extended system.
First, Equation (11.3) is rewritten as Equation (11.7):

$$\bar{P}(s) = \frac{y(s)}{u(s)} = \frac{\beta_1 s^{\bar{n}-1} + \cdots + \beta_{\bar{n}}}{s^{\bar{n}} + \alpha_1 s^{\bar{n}-1} + \cdots + \alpha_{\bar{n}}}$$

$$= \frac{\beta_1 s^{\bar{n}-1} + \cdots + \beta_{\bar{n}}}{s^{\bar{n}} + (k_1 - l_1)s^{\bar{n}-1} + \cdots + (k_{\bar{n}} - l_{\bar{n}})} \quad (11.7)$$

Let $l_i = k_i - \alpha_i$, and choose k_i so that $k(s) = s^{\bar{n}} + k_1 s^{\bar{n}-1} + \cdots + k_{\bar{n}}$ is stable. Next, divide the denominator and numerator of Equation (11.7) by $k(s)$ and organize it by multiplying $y(s)$ and $u(s)$ with cross-multiplication. Then, we obtain

$$\frac{k(s) - (l_1 s^{\bar{n}-1} + \cdots + l_{\bar{n}})}{k(s)} y(s) = \frac{\beta_1 s^{\bar{n}-1} + \cdots + \beta_{\bar{n}}}{k(s)} u(s) \quad (11.8)$$

through which Equation (11.9) is obtained:

$$y(s) = \frac{l_1 s^{\bar{n}-1} + \cdots + l_{\bar{n}}}{k(s)} y(s) + \frac{\beta_1 s^{\bar{n}-1} + \cdots + \beta_{\bar{n}}}{k(s)} u(s) \quad (11.9)$$

where $f_{yi}(s) = \frac{s^{\bar{n}-i} y(s)}{k(s)}$, and $f_{ui}(s) = \frac{s^{\bar{n}-i} u(s)}{k(s)}$ are defined as $y(s) = \Sigma_{i=1}^{\bar{n}} \{l_i f_{yi}(s) + \beta_i f_{ui}(s)\}$. In addition,

$$f_y^T(s) = \begin{bmatrix} f_{y1}(s) & f_{y2}(s) & \cdots & f_{y\bar{n}}(s) \end{bmatrix} \in R^{1 \times \bar{n}}.$$

Let $f_u^T(s) = \begin{bmatrix} f_{u1}(s) & f_{u2}(s) & \cdots & f_{u\bar{n}}(s) \end{bmatrix} \in R^{1 \times \bar{n}}$ and denote $y(s)$ by Equation (11.10):

$$y(s) = l^T f_y(s) + \beta^T f_u(s) \quad (11.10)$$

where $l^T = [l_1, l_2, \ldots, l_{\bar{n}}]$, $\beta^T = [\beta_1, \beta_2, \ldots, \beta_{\bar{n}}]$, and $K = \begin{bmatrix} -k_1 & 1 & 0 & \cdots & 0 \\ -k_2 & 0 & 1 & 0 & 0 \\ \vdots & \vdots & & \ddots & 1 \\ -k_{\bar{n}} & 0 & 0 & \cdots & 0 \end{bmatrix}$.

To create $(sI - K^T) \cdot f_y(s)$,

$$(sI - K^T) f_y(s)$$

$$= \begin{bmatrix} s + k_1 & k_2 & \cdots & k_{\bar{n}} \\ -1 & s & 0 & 0 \\ \vdots & \ddots & \ddots & 0 \\ 0 & 0 & -1 & s \end{bmatrix} \cdot \begin{bmatrix} f_{y1} \\ f_{y2} \\ \vdots \\ f_{y\bar{n}} \end{bmatrix} = \begin{bmatrix} \bar{k}(s) \\ -f_{y1} + s f_{y2} \\ \vdots \\ -f_{y\bar{n}-1} + s f_{y\bar{n}} \end{bmatrix}$$

where $\bar{k}(s) = (s + k_1) f_{y1} + k_2 f_{y2} + \cdots + k_{\bar{n}} f_{y\bar{n}}$. Using the defining equations for f_{yi} and f_{ui} on the right side, the second line becomes $-f_{y1} + s f_{y2} = 0$, and everything after the third line is also 0. Additionally, the $\overline{k}(s) = (s + k_1)\frac{s^{\overline{n}-1}y(s)}{k(s)} +$

$k_2\frac{s^{\overline{n}-2}y(s)}{k(s)} + \cdots = \frac{s^{\overline{n}}y(s)}{k(s)} + \frac{k_1 s^{\overline{n}-1}y(s)}{k(s)} + \cdots = y(s)$.

With this result and combining $\overline{C}^T y(s) = \begin{bmatrix} y(s) \\ 0 \\ \vdots \\ 0 \end{bmatrix}$, we obtain $f_y(s) = (sI - K^T)^{-1}\overline{C}^T y(s)$.

Similarly, we obtain $f_u(s) = (sI - K^T)^{-1}\overline{C}^T u(s)$. Here, applying the inverse Laplace transform with 0 as the initial values for $y(t)$ and $u(t)$ yields Equations (11.11) and (11.12) in the time domain:

$$\dot{f}_y(t) = K^T f_y(t) + \overline{C}^T y(t) \tag{11.11}$$

$$\dot{f}_u(t) = K^T f_u(t) + \overline{C}^T u(t) \tag{11.12}$$

Now, combining Equations (11.11), (11.12), and (11.10), we can draw Figure 11.2. In the figure, the dashed boxes correspond to Equations (11.11) and (11.12), which we call the state variable filter. The inside of the double-dotted box corresponds to Equation (11.10). Inside the single-dotted box are $l_i = k_i - \alpha_i$ and β_i, which we say are unknown parameters. In other words, Figure 11.2, upper part, is the control

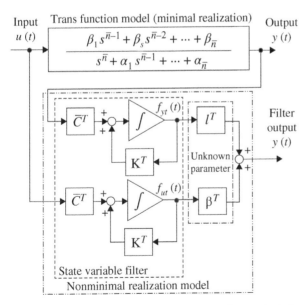

Figure 11.2 Nonminimal realization model of controlled plant with disturbance.

plant represented by the minimum realization, and the lower part is the model of the nonminimal realization form derived to construct the adaptive observer. It is important to note that both models produce the same output if the unknown parameters are correctly obtained.

This can be expressed as Equation (11.13) by letting the unknown parameters be $\phi^T = [l^T, \beta^T]$ and $\xi^T(t) = [f_y^T(t), f_u^T(t)]$:

$$y(t) = l^T f_y(t) + \beta^T f_u(t) = [l^T, \beta^T] \begin{bmatrix} f_y(t) \\ f_u(t) \end{bmatrix} = \phi^T \xi(t) \tag{11.13}$$

Example 11.3.1 Let us show an example simulation corresponding to Figure 11.2 and check that the nonminimal realization model is equivalent to the original minimal realization model. The program is shown in List 11.1, and Figures 11.3 and 11.4 show the Simulink model and output waveform, respectively.

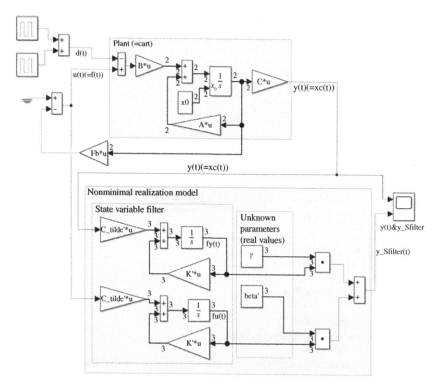

Figure 11.3 Simulink model of nonminimal realization model.

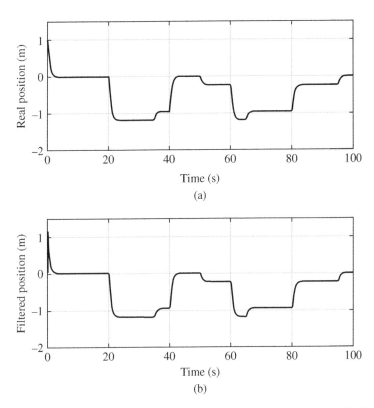

Figure 11.4   Comparison of positional waveforms for nonminimal realization models.

List 11.1:   Example of evaluating a nonminimal realization model.

```
1   %% Physical paremeters
2   m=2.0;c=0.30;k=0.050;
3
4   %% Create extended space model
5   A=[0, 1; −k/m,−c/m];B=[0;1/m];C=[1,0];D=0;[n,~]=size(B);
6   A_bar=[A,−B;zeros(1,3)];B_bar=[B;0];C_bar=[C,0];n_bar=n+1;
7   %
8   lambda1=eig(A_bar);x=poly(lambda1);% Compute eigenvalues and obtain
9   % polynomial coefficients
10  a3=x(2);a2=x(3);a1=x(4); % Define s^3+a3*s^2+a2*s+a1
11  W=[a2,a3,1;a3,1,0;1,0,0]; % W matrix for creating observable canonical systems
12  Uo=[C_bar;C_bar*A_bar;C_bar*A_bar^2];S=W*Uo; % Creation of transformation matrix S
13  A_t=S*A_bar*inv(S);B_t=S*B_bar;C_t=C_bar*inv(S);
14  % Rearrange the order to create an observable canonical system for
15  % adaptive observers
16  S2=[0,0,1;0,1,0;1,0,0];
17  A_tilde=S2*A_t*inv(S2);B_tilde=S2*B_t;C_tilde=C_t*inv(S2);
18
```

```
19   %% State variable filter design
20   g=[1;zeros(n_bar−2,1)];
21   L=[zeros(n_bar−2,1),eye(n_bar−2);0,zeros(1,n_bar−2)];
22     pole_of_k=[−6,−9,−12]; % Specification of stable k(s) eigenvalues
23   k_temp=poly(pole_of_k);k=k_temp(2:end); % Creation of k−vector
24   K=[−k',[g';L]]; % Creating a K−matrix for a state variable filter
25
26   %% Correct values of alpha and beta
27   [num,den]=ss2tf(A_bar,B_bar,C_bar,0);
28   alpha=(den(2:end))',l=−alpha+k';beta=(num(2:end))',
29
30   %% Control systen design
31   Q=diag([2.5e2,1e2]);R=1;Fb=lqr(A,B,Q,R);
32
33   %% Simulation
34   x0=[1;0];
35   tend=100;dis_0=0;dis=15;t_dis=20;
36   open_system('sim_Figure_11_3_and_11_4_nonminimal_realization_model');
37   set_param('sim_Figure_11_3_and_11_4_nonminimal_realization_model','WideLines','on'); % Draws
         non−scalar lines thicker.
38   set_param('sim_Figure_11_3_and_11_4_nonminimal_realization_model','ShowLineDimensions','on')
39   z=sim('sim_Figure_11_3_and_11_4_nonminimal_realization_model');
```

At this point, the unknown parameters are not yet estimated, but the actual values of l and β are used. The original model under control (minimal realization) and the output of the designed nonminimal realization model should match, and there is a good agreement for Figure 11.4. However, the degree of agreement varies depending on the design of the stable eigenvalues of the K matrix, and choosing a slower valence closer to the imaginary axis results in larger delay and error. In other words, it should be recognized that the nonminimal realization model used here has dynamics not present in the minimal realization model.

11.4 Design of Kreisselmeier-Type Adaptive Disturbance Observer

We rewrite the above derived observable canonical system to Equations (11.14) and (11.15):

$$\dot{\bar{z}}(t) = K\bar{z}(t) + (-\alpha + k)y(t) + \tilde{B}u(t)$$

$$= \begin{bmatrix} -k_1 & 1 & 0 & \cdots & 0 \\ -k_2 & 0 & 1 & 0 & 0 \\ \vdots & \vdots & \ddots & \ddots & 1 \\ -k_{\bar{n}} & 0 & \cdots & 0 & 0 \end{bmatrix} \bar{z}(t) + \left(-\begin{bmatrix} \alpha_1 \\ \alpha_2 \\ \vdots \\ \alpha_{\bar{n}} \end{bmatrix} + \begin{bmatrix} k_1 \\ k_2 \\ \vdots \\ k_{\bar{n}} \end{bmatrix} \right) y(t)$$

$$+ \begin{bmatrix} \beta_1 & \beta_2 & \cdots & \beta_{\bar{n}} \end{bmatrix}^T u(t) \tag{11.14}$$

$$y(t) = \tilde{C}\bar{z}(t) \tag{11.15}$$

This special **observable canonical form** is used to estimate the state and output variables. Now, assuming that the control plants are represented by Equations (11.14) and (11.15), we define the following:

$$\dot{R}_y(t) = K R_y(t) + I_{\bar{n}} y(t) \tag{11.16}$$

$$\dot{R}_u(t) = K R_u(t) + I_{\bar{n}} u(t) \tag{11.17}$$

Although $R_y(t) \in R^{\bar{n} \times \bar{n}}$ and $R_u(t) \in R^{\bar{n} \times \bar{n}}$ are not described specifically at this stage, assume the following relation with the state variable filter $f_y(t) = R_y^T(t)\tilde{C}^T$ and $f_u(t) = R_u^T(t)\tilde{C}$, and let $R_y(0) = 0$, $R_u(0) = 0$. The Laplace transform of Equation (11.14) gives $s\bar{z}(s) - \bar{z}(0) = K\bar{z}(s) + (-\alpha + k)y(s) + \tilde{B}u(s)$.

Then, Equations (11.16) and (11.17) are also Laplace transformed to the equations $(sI_{\bar{n}} - K)R_y(s) = I_{\bar{n}}y(s)$ and $(sI_{\bar{n}} - K)R_u(s) = I_{\bar{n}}u(s)$, and we get $\bar{z}(s) = R_y(s)(\alpha + k) + R_u(s)\tilde{B} + (sI_n - K)^{-1}\bar{z}(0)$. Finally, we then apply an inverse Laplace transform to obtain Equation (11.18)[5]:

$$\bar{z}(t) = R_y(t)(-\alpha + k) + R_u(t)\tilde{B} + e^{Kt}\bar{z}(0) \tag{11.18}$$

Since $y(t) = \tilde{C}\bar{z}(t)$, we obtain

$$\begin{aligned} y(t) &= \tilde{C}\bar{z}(t) = \tilde{C}R_y(t)(-\alpha + k) + \tilde{C}R_u(t)\tilde{B} + \tilde{C}e^{Kt}\bar{z}(0) \\ &= f_y^T(t)(-\alpha + k) + f_u^T(t)\tilde{B} + \tilde{C}e^{Kt}\bar{z}(0) \end{aligned} \tag{11.19}$$

Then, using $f_y(t) = R_y^T(t)\tilde{C}^T$, we get $\begin{bmatrix} \tilde{C} \\ \tilde{C}K \\ \vdots \\ \tilde{C}K^n \end{bmatrix} R_y(t) = \begin{bmatrix} f_y^T(t) \\ f_y^T(t)K \\ \vdots \\ f_y^T(t)K^n \end{bmatrix}$, and express $R_y(t)$ as

in Equation (11.20). For $R_u(t)$, similarly replace the subscript y with u.

$$R_y(t) = \begin{bmatrix} \tilde{C} \\ \tilde{C}K \\ \vdots \\ \tilde{C}K^n \end{bmatrix}^{-1} \begin{bmatrix} f_y^T(t) \\ f_y^T(t)K \\ \vdots \\ f_y^T(t)K^n \end{bmatrix} \tag{11.20}$$

where $\phi^T(t) = [l^T(t), \beta^T(t)]^T = [(-\hat{\alpha}(t) + k)^T, \beta^T(t)]^T$, and letting $\xi^T(t) = [f_y^T(t), f_u^T(t)]^T$ and $f_{y0}(t) = \tilde{C}e^{Kt}\bar{z}(0)$, we obtain Equation (11.21):

$$y(t) = \phi^T \xi(t) + f_{y0}(t) \tag{11.21}$$

Determine the structure of the adaptive disturbance observer based on the previous preparations.

$$\hat{\bar{z}}(t) = R_y(t)\{-\hat{\alpha}(t) + k\} + R_u(t)\hat{\beta}(t) \tag{11.22}$$

---

5 The order was rearranged to account for the dimensions.

$\hat{\alpha}$ and $\hat{\beta}(= \tilde{B})$ are the unknown parameter estimates, and the output is defined as $\hat{y}(t) = \hat{\phi}^T(t)\xi(t)$.[6] Next, let the error vector be $\epsilon(t) = y(t) - \hat{y}(t)$ for $\hat{y}(t) = \hat{\phi}^T\xi(t)$, and we obtain the unknown parameter $\hat{\phi}$ with $\xi(t)$.

Among the various adaptive laws, we use Equations (11.23) and (11.24), which are the basic weighted least-squares adaptive identification laws:

$$\dot{\hat{\phi}}(t) = -\Gamma(t)^{-1}\xi(t)\epsilon(t) \tag{11.23}$$

$$\dot{\Gamma}(t) = -\lambda\Gamma(t) + \xi^T(t)\xi(t) \tag{11.24}$$

By the way, for the adaptation law to hold, the error model $W(t)$ in the error equation must be a **strictly positive real** [5]. Since the output is $Y(t) = \phi^T\xi(t)$ in Equation (11.13), and the output estimating equation is $\hat{y}(t) = \hat{\phi}^T\xi(t)$, the error equation $\epsilon(t) = y(t) - \hat{y}(t) = (\phi^T - hat\phi^T)\xi(t) = \Theta^T\xi(t) = W(t)\Theta^T\xi(t)$ and $W(t) = 1$. Since this $W(t)$ is the simplest strict positive real, we can use the adaptation law.

Example 11.4.1 List 11.2 shows an example of a control system with an adaptive disturbance observer using a cart model and Figure 11.5 shows the corresponding Simulink model. List 11.3 shows the function of the 'Adaptive Estimator' within Figure 11.5. List 11.4 shows the state feedback controller 'State_FB' within Figure 11.5. The state feedback controller is designed using the estimated parameters and the pole assignment method. Moreover, Figure 11.6 (a) shows the contents in the 'State_variable filter' and (b) shows the contents of the subsystem 'Calc. of inv_Gmma' Subsystem in Figure 11.5. Finally, The simulation results are shown in Figs.11.7 and 11.8.

List 11.2: Example of control system with adaptive disturbance observer.

```
1    %% Physical parameters
2    m=2.0;c=0.30;k=0.050;
3
4    %% Create extended space model
5    A=[0, 1; −k/m ,−c/m];B=[0;1/m];C=[1,0];D=0;[n,~]=size(B);
6    A_bar=[A,−B;zeros(1,3)];B_bar=[B;0];C_bar=[C,0];n_bar=n+1;
7    %
8    lambda1=eig(A_bar);x=poly(lambda1);% Compute eigenvalues and obtain
9    % polynomial coefficients
10   a3=x(2);a2=x(3);a1=x(4); % Define s^3+a3*s^2+a2*s+a1
11   W=[a2,a3,1;a3,1,0;1,0,0]; % W matrix for creating observable canonical systems
12   Uo=[C_bar;C_bar*A_bar;C_bar*A_bar^2];S=W*Uo; % Creation of transformation matrix S
13   A_t=S*A_bar*inv(S);B_t=S*B_bar;C_t=C_bar*inv(S);
```

---

6 $\hat{f}_{y0}(t) = 0$ is assumed.

```
14   % Rearrange the order to create an observable canonical system for
15   % adaptive observers
16   S2=[0,0,1;0,1,0;1,0,0];
17   A_tilde=S2*A_t*inv(S2);B_tilde=S2*B_t;C_tilde=C_t*inv(S2);
18
19   %% State variable flter design
20   g=[1;zeros(n_bar−2,1)];
21   L=[zeros(n_bar−2,1),eye(n_bar−2);0,zeros(1,n_bar−2)];
22   pole_k=[−3,−4,−2];kk_temp=poly(pole_k);kk=(kk_temp(2:4))';
23   K=[−kk,[g';L]] % Create K matrix of state variable filter
24
25   %% Kreisselmeierfilter design
26   C_tilde_K_matrix=[C_tilde;C_tilde*K;C_tilde*K^2];
27   inv_C_tilde_K_matrix=inv(C_tilde_K_matrix);
28
29   %% Calculate exact alpha and beta values for evaluation
30   [num,den]=ss2tf(A_bar,B_bar,C_bar,0);
31   alpha=(den(2:end))';l=−alpha+kk;beta=(num(2:end))';
32   alpha_beta=[alpha;beta];l_beta=[l;beta];
33
34   %% Setting gain of adaptive law
35   lambda=1.5e−1;gamma0=5*eye(6);
36
37   %% Preparation of Controller design for calculation
38   % with pole assignment method performed in MATLAB function
39   pole=[−2,−3];h=poly(pole);h1=h(3);h2=h(2);
40
41   %% Simulation
42   x0=[0;0];
43   tend=200;dis_0=5.0;dis=5;t_dis=20;T_ratio=50;
44   open_system('sim_Figure_11_5_6_7_8_adaptive_DOB_and_Regulator');
45   set_param('sim_Figure_11_5_6_7_8_adaptive_DOB_and_Regulator','WideLines','on');
46   set_param('sim_Figure_11_5_6_7_8_adaptive_DOB_and_Regulator','ShowLineDimensions','on')
47   z=sim('sim_Figure_11_5_6_7_8_adaptive_DOB_and_Regulator');
```

List 11.3:   Adaptive estimator.

```
1   function [y_hat,ab_hat,z_bar_hat]=Estimator(Phi,Ry,Ru,C_tilde,l,beta, kk)
2   %.
3   l=Phi(1:3);beta=Phi(4:6);
4   z_bar_hat=Ry*l+Ru*beta;
5   y_hat=C_tilde*z_bar_hat;
6   %(1:6)
7   alpha=−l+kk;% l=−alpha+kk
8   ab_hat=[alpha;beta];
```

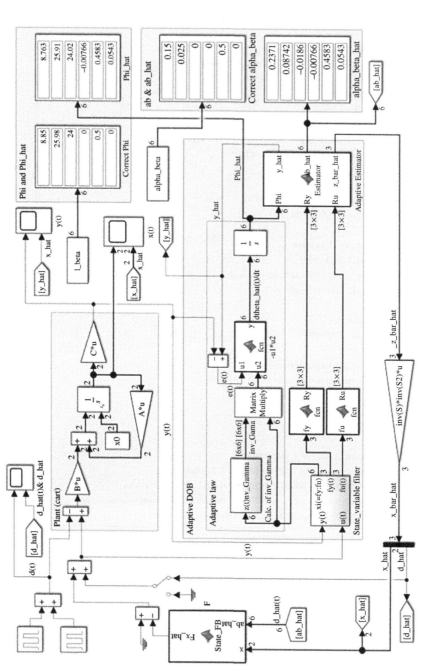

Figure 11.5 Simulink model of an adaptive disturbance observer control system.

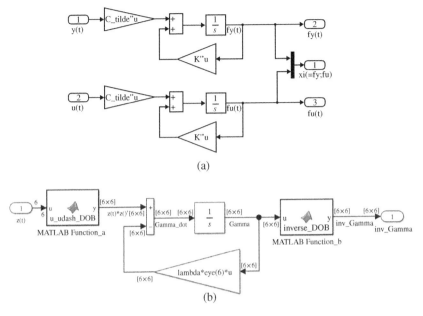

Figure 11.6   Subsystems for adaptive disturbance observer. (a) Block diagram of the state variable filter and (b) adaptive law.

List 11.4:   Adaptive control law adaptive controller.

```
1  function Fx_hat = State_FB(x,ab_hat,h1,h2)
2  Calculation of state feedback control inputs using the pole assignment method
3  a1=ab_hat(2);a2=ab_hat(3);AA=[0,1;-a2,-a1];BB=[0;1];
4  Uc=[BB,AA*BB];W=[a1,1;1,0];T=Uc*W;inv_T=inv(T);
5  Fbb=[h2-a2,h1-a1]*inv_T; Fx_hat=Fbb*x;
```

The content of the MATLAB function_a in Figure 11.6b is only $y = u * u'$, and that of the MATLAB function_b is $y = inv(u)$. The content of $-u1 * u2$ in fcn of Figure 11.5 is also $y = -u1 * u2$. The contents of the blocks "input fy output Ry" and "input fu output Ru" in the MATLAB function fcn are as in Lists 11.5 and 11.6.

List 11.5:   Calculation of Ry.

```
1  function Ry = fcn(fy,inv_C_tilde_K_matrix, K)
2  fyK=[fy';fy'*K;fy'*K^2];
3  Ry=inv_C_tilde_K_matrix*fyK;
```

List 11.6:   Calculation of Ru.

```
1  function Ru = fcn(fu,inv_C_tilde_K_matrix, K)
2  fuK=[fu';fu'*K;fu'*K^2];
3  Ru=inv_C_tilde_K_matrix*fuK;
```

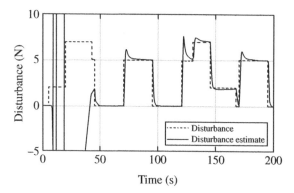

Figure 11.7 Disturbance estimation for adaptive DOB combined control system.

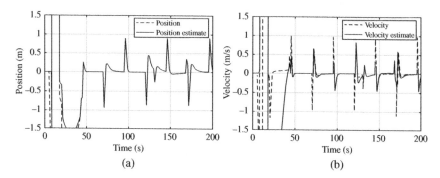

Figure 11.8 Simulation results of a control system with an adaptive disturbance observer. (a)Estimated waveform of position and (b) estimated waveform of velocity.

Figure 11.6a is equivalent to the state variable filter of Figure 11.3.

The disturbance estimate $\hat{d}(t)$ in Figure 11.7 follows the real disturbance $d(t)$. The disadvantages are that a large error occurs until the initial convergence, and that, it is difficult to adjust the control parameters. Therefore, it is necessary to consider using other adaptive control methods with better performance. However, the waveform in Figure 11.8 is greatly disturbed at the beginning of the disturbance but gradually stabilizes, and the position estimation waveform in (a) accurately follows the actual position in response to subsequent disturbance changes. The velocity waveforms in Figure 11.8b show a slight error.

References

1 Akira Shimada, Danai Phaoharuhansa: An Adaptive Disturbance Observer Design on Motion Control Systems, Proceedings of the 2012 International Conference on Advanced Mechatronic Systems, 320–322, 2012.

2 Gerhard Kreisselmeir: Adaptive observer with exponential rate of convergence, IEEE Transactions on Automatic Control, Vol. 22, No. 1, 2–8, 1977.

3 Seizo Fujii, Naoki Mizuno: Multivariable discrete model reference adaptive control using an autoregressive model with dead time of the plant and its application, Transactions of the Society of Instrument and Control Engineers (SICE), Vol. E-2, No. 1, 178–186, 2002.

4 Karl Journal Åstroem, Bjoern Wittenmark: Adaptive Control, Second Edition, Dover Publications, 2008.

5 Petros Ioannou, Jing Sun: Robust Adaptive Control, Dover Publications, 2012.

12

Methods for Measuring and Estimating Velocities

When implementing a velocity-observing disturbance observer, a high-performance **velocity measurement and estimation** method that considers noise is required. This chapter is based on the literature of Prof. Tsuji and Mr. Nagatomi et al. and summarizes the velocity measurement and estimation methods with their cooperation [1–3].[1]

12.1 Importance of Velocity Measurement

The rotary encoder is a typical sensor used to measure the rotation angle. The most critical specification item of a rotary encoder is its resolution, which depends on the number of pulses generated per revolution. The position resolution Q_p of the encoder with P pulse/revolution (PPR) is expressed in Equation (12.1):

$$Q_p = \frac{2\pi}{P} \tag{12.1}$$

The quantization error is the error that occurs when converting a continuous signal into a discrete signal. The encoder is a sensor that counts the number of pulses generated to obtain position information, and the output value is discrete. Q_p is the maximum quantization error generated by the encoder. For example, an encoder with 2000 PPR will have a quantization error of up to $2\pi/2000$.

---

[1] However, it is difficult clearly to distinguish between "velocity measurement" and "velocity estimation." In this book, the methods with motion models refer to as velocity estimation methods.

Disturbance Observer for Advanced Motion Control with MATLAB/Simulink, First Edition. Akira Shimada.
© 2023 The Institute of Electrical and Electronics Engineers, Inc. Published 2023 by John Wiley & Sons, Inc.
Companion website: www.wiley.com/go/disturbanceobserver

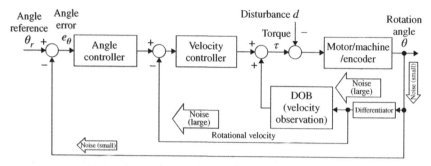

Figure 12.1 Noise propagation in angle control systems.

Velocity resolution[2] is an essential element for determining the performance of a control system. Therefore, using a typical control system with position and velocity feedback loops as an example, we will explain with Figure 12.1.[3] The arrows indicate noise generated in the feedback loop. The position information obtained from the encoder contains noise due to quantization errors. The velocity information is obtained by differentiating the angle information, but the derivative operation amplifies the noise in the angle information. As the simplest example of a differential operation, the expression for the backward difference is shown in Equation (12.2):

$$\overline{\omega}(k) = \frac{\overline{\theta}(k) - \overline{\theta}(k-1)}{T} \tag{12.2}$$

$\overline{\theta}$ is the measured angle value, $\overline{\omega}$ is the calculated angular velocity, T is the sampling period (=control period), and k is the number of sample points. The superscripted bar indicates the measured value. Since $\overline{\theta}$ has a maximum error of Q_p, the velocity resolution Q_v is expressed using Equation (12.3):

$$Q_v = \frac{Q_p}{T} = \frac{2\pi}{PT} \tag{12.3}$$

Thus, the shorter the sampling period, the greater the velocity noise. A smoothing process is often applied since velocity feedback noise degrades control performance. However, increasing the time constant during the smoothing process reduces the noise but increases the time delay, whereas reducing the time delay increases the noise. Therefore, a good balance between noise and time delay should be selected, but a better algorithm can reduce the noise of velocity measurement even with the same time delay.

---

2 In this book, the maximum error in a theoretically derived velocity value is called the velocity resolution. Velocity resolution varies with the encoder's performance and the algorithm of velocity calculation.

3 The current minor loop in the motor driver also exists but will be omitted since it does not involve an encoder.

12.2 Velocity Measurement and Estimation Methods

12.2.1 Pseudo-derivative

In actual systems, it is common to combine a **low-pass filter** (LPF) to deal with the problem of amplification by differential operations of high-frequency noise.

Although the ideal derivative is called the perfect derivative, Equation (12.4) is often used, which is called a pseudo-derivative[4]:

$$\overline{\omega}(s) = \frac{gs}{s+g}\overline{\theta}(s) \tag{12.4}$$

Let g be the cutoff frequency of the filter. Equation (12.4) can also be rewritten as Equation (12.5):

$$\overline{\omega}(s) = \frac{g}{s+g} \cdot s \cdot \overline{\theta}(s) \tag{12.5}$$

Moreover, Equation (12.4) converted to a digital system is shown in Equation (12.6)[5]:

$$\overline{\omega}(z) = \frac{g(1-z^{-1})/T}{g+(1-z^{-1})/T}\overline{\theta}(z) \tag{12.6}$$

Figure 12.2 represents a step response comparison of the exact and pseudo-derivatives. In the exact derivative, the derivative of the step response yields an infinite pulse-like response value. In the presence of noise, such as in the step

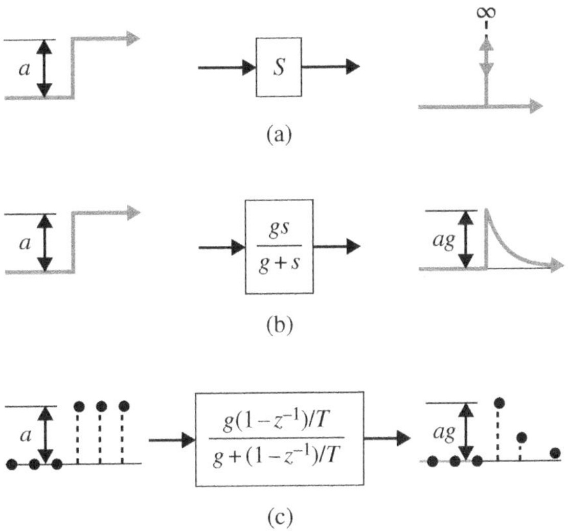

Figure 12.2 Comparison of step response between exact- and pseudo-derivative. (a) Ideal derivative, (b) lagged derivative, and (c) lagged derivative in digital system.

4 It is also called an incomplete derivative (lagged derivative), and **pseudo-derivative** is often used.
5 There are various methods for converting a continuous system to a digital system, but in this example, we used **backward difference** $s = \frac{1-z^{-1}}{T}$.

response, which fluctuates in value over a short period, the derivative becomes vast and amplifies the noise. In pseudo-differentiation, the maximum value of the derivative is smaller than that of the exact derivative, and positive derivative values remain after the step response. Additionally, integrating the derivative will yield the width, a, of the step response.

12.2.2 Counting and Timekeeping Methods

In mechatronics systems equipped with rotary encoders, the **counting** and **timing calculation method** are often used. These are based on obtaining the velocity by processing the pulse signal sent from "the encoder" as shown in Figure 12.3, instead of obtaining the velocity by differentiating the position with time. The counting method involves counting the number of pulses generated each time. The number of pulses refers to the number of pulse signals generated by the encoder during each control period. In an incremental encoder, a positive number of pulse signals are generated from phases A and B in both positive and negative rotation directions, but the pulses generated in the negative direction of rotation are counted as a negative number and called the pulse count. Thus, the number of pulses is an integer proportional to the magnitude of the displacement.[6]

The displacement between the sample points can be obtained from the number of pulses generated, and the velocity can be derived by dividing by the control period. The concept is shown in Figure 12.4.

The m_e denotes the number of pulses generated in one control cycle. The measured velocity is in Equation (12.7).

$$\overline{\omega}(i) = \frac{2\pi m_e}{PT} \tag{12.7}$$

Letting $\overline{\theta}(i)$ be the positional measurement of the ith sample

$$\frac{2\pi m_e}{P} = \overline{\theta}(i) - \overline{\theta}(i-1) \tag{12.8}$$

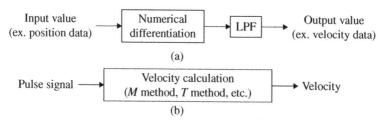

(a)

(b)

Figure 12.3 Concepts of (a) general differential operations and (b) velocity operations with encoders.

---

6 However, an absolute encoder does not generate a pulse signal, but the equivalent of the number of pulses of an incremental encoder can be obtained by taking the difference between two sample points.

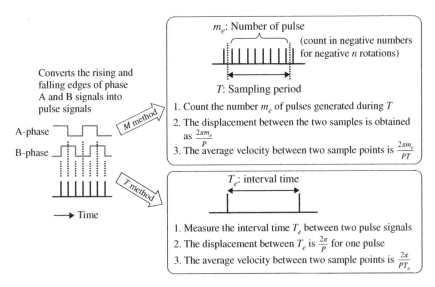

Figure 12.4 Conceptual diagram of M and T methods.

Substituting this into the expression (12.7) yields

$$\overline{\omega}(i) = \frac{\overline{\theta}(i) - \overline{\theta}(i-1)}{T} \tag{12.9}$$

Therefore, this method is equivalent to the result of first-order precision backward differentiation of angle information. Furthermore, adding an LPF to the counting method to reduce noise is equivalent to pseudo-differentiation. The velocity resolution Q_v without smoothing is as shown in Equation (12.3), and the measurement time T_m is equal to the sampling period T because the velocity is derived at each sampling period (=control period)[7]:

$$T_m = T \tag{12.10}$$

The counting method has the advantage of a short measurement time, but its measurement accuracy is low, which is a problem at low velocities. Meanwhile, the timekeeping method improves velocity measurement accuracy when noise is a problem in the low-velocity range. The timekeeping method is a method to measure the time interval between two consecutive pulse signals. The velocity value is calculated by dividing the angle $\frac{2\pi}{P}$ for one pulse by the time between pulses. In Figure 12.4, T_e represents the time between pulses and is measured as $T_e = m_s T$, a constant multiple of the sampling period T. The M_s is the number of

---

7 In this chapter, we denote "sampling period" rather than "control period" when we wish to emphasize that it is the period during which sensor information is acquired.

sampling elapsed between pulses. Therefore, the time between pulses T_e involves an error of up to T. The velocity value is derived from Equation (12.11):

$$\overline{\omega} = \frac{2\pi}{m_s PT} \tag{12.11}$$

The velocity resolution Q_v and measurement time T_m are expressed using the following equations:

$$Q_v = \frac{2\pi}{m_s(m_s - 1)PT} \tag{12.12}$$

$$T_m = m_s T \tag{12.13}$$

Since the time interval is measured when the pulse signal is generated, the measurement time T_m varies depending on the timing at which the pulse signal is generated. While the measurement time is m_s times longer, the velocity resolution Q_v is $\frac{1}{m_s(m_s-1)}$. Although the timekeeping method has a high-velocity resolution, it also has its problems; the measurement time is variable.

The measurement time T_m becomes very long if the shaft rotates at near-zero velocity, or if the encoder resolution is low. The time delay degrades the control system's performance. An upper limit on the measurement time should be set to avoid performance degradation. It is necessary to force the velocity derivation when it is exceeded.

The next point is that the velocity range that can be measured is limited. The higher the rotation velocity, the shorter the pulse time interval T_e, and T_e is less than T. Measuring the time between pulses becomes impossible when the time between pulses is short. Although the units of the designed sampling period T are the base of the time interval, the use of a dedicated timer circuit makes the accuracy of the interval time correct and the velocity range also wider. It is relatively easy to implement since many recent DSPs are equipped with a timer unit as standard equipment. The counting method changes the number of pulses m_e with velocity, and the measurement time T_m is constant. However, in the timekeeping method, the measurement time T_m varies with velocity, and the number of pulses m_e is always 1.

12.2.3 M/T Method

Since the counting method has low accuracy, and the timekeeping method cannot measure in the high-velocity range, we introduce the **M/T method** as a high-precision velocity measurement method in Figure 12.5 [4].[8]

---

8 Although there are the Constant Sample-time Digital Tachometer (CSDT) [5] and the CONSTANT ELAPSED TIME (CET) [6] in addition to the M/T method, the M/T method is represented in this book.

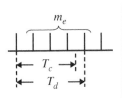

1. Count the number of pulses until the minimum measurement time T_c elapses

2. After T_c elapses, the first pulse is detected and its time T_d is recorded

3. If the number of pulses generated during T_d is m_e, the velocity is obtained as $\bar{\omega} = \frac{2\pi m_e}{PT_d}$

Figure 12.5 Algorithm of the M/T method.

In the M/T method, the minimum measurement time T_c is set in advance, and after T_c has elapsed, the time T_d until the first pulse signal is detected is measured. Then, the number of pulses generated during T_d is calculated as m_e, and the velocity is derived using Equation (12.14):

$$\bar{\omega} = \frac{2\pi m_e}{PT_d} \tag{12.14}$$

This algorithm solves the problem of timekeeping methods where the time interval of the pulse signal cannot be measured in the high-velocity range. The velocity resolution Q_v and measurement time T_m are expressed by the following equations:

$$Q_v = \frac{2\pi m_e}{m_s(m_s - 1)PT} \tag{12.15}$$

$$T_m = T_d = m_s T \tag{12.16}$$

At low velocities with velocity $\omega < \frac{2\pi}{PT_c}$, no pulse signal is generated within the times T_c, and $m_e = 1$ without fail. Substituting this into Equation (12.15) becomes equal to Equation (12.12), representing the timekeeping method's velocity resolution. In this case, the measurement time T_m agrees with that of the timekeeping method. In other words, the M/T method has the same effect as the timekeeping method in the low-velocity range. The M/T method is superior to the timekeeping method because it can measure velocity even at high velocities. As the velocity increases, the value of m_e increases, and the error increases gradually. Higher resolution than the counting method can be obtained unless the axis rotates at a very high velocity. By measuring T_d with an additional timer unit with a short sampling time, the M/T method is implemented in many high-precision control systems because it can achieve high-precision measurement over a wide velocity range.

12.2.4 Synchronous Counting Method

We introduce the **synchronous counting method** for high-precision measurements in the high-velocity range without additional timer units [1, 2].

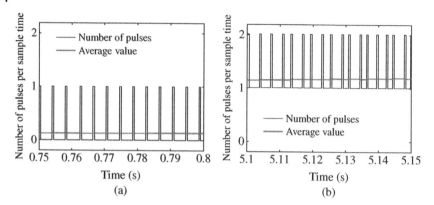

Figure 12.6 Response of the number of pulses in the experiment. (a) Low-velocity range and (b) high-velocity range.

The synchronous counting method focuses on a typical pattern in the response of the number of pulses and performs high-precision measurements in the high-velocity range. Figure 12.6 compares the response of the number of pulses in the low- and high-velocity ranges and the average value. Both figures show that the pulse number patterns are similar in the low- and high-velocity ranges. The pulse count is basically zero, and samples with a pulse count of one occasionally appear. Next, in the high-velocity region, the pulse count changes in a pattern similar to that in the low-velocity region. In the high-velocity region, n is the number of pulses, with occasional samples in which the number of pulses changes to $n + 1$ or $n - 1$. The change in the number of pulses is called pulse variation. Figure 12.7 shows the algorithm for finding the average velocity synchronized with the pulse variation. The synchronous counting method extends the timekeeping algorithm to the total velocity range and differs from the timekeeping method since it synchronizes the pulse variations. In contrast, the timekeeping method derives the velocity synchronously with the pulse. The velocity resolution Q_v and measurement time T_m are expressed in Equations (12.17) and (12.18):

$$Q_v = \frac{2\pi}{m_s(m_s - 1)PT} \tag{12.17}$$

$$T_m = m_s T \tag{12.18}$$

Unlike the M/T method, the velocity resolution Q_v does not depend on the number of pulses m_e. The accuracy does not deteriorate even at high velocity. Highly accurate measurements can be achieved with the M/T method in most velocity ranges if there is an additional timer circuit.

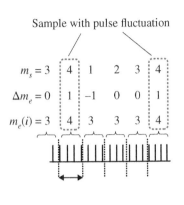

Sample with pulse fluctuation

$m_s = 3$

$\Delta m_e = 0$

$m_e(i) = 3$

1. Count the number of pulses m_e for each sampling period T_s
2. Derive the number of pulse fluctuation
$$\Delta m_e = m_e(k) - m_e(k-1)$$
3. If $\Delta m_e = 0$ or $m_s = 1$, return to 1. Otherwise, if $\Delta m_e \neq 0$ and $m_s > 1$, then a pulse fluctuation is considered to have occurred and go to 4.
(m_s: number of samples elapsed since the last pulse variation)
4. The velocity is obtained as
$$\bar{\omega} = \frac{2\pi \Sigma_{j=0}^{m_s-1} m_e(k-j)}{P m_s T_s}$$

Figure 12.7 Algorithm for synchronous counting method.

12.2.5 Instantaneous Velocity Observer

The velocity observer uses the servo motor's current information to estimate the velocity between the encoder sampling times. The principle is shown in Figure 12.8. The upper half of the figure shows the actual motor, and the lower half shows the observer based on the current and estimated disturbance values. The controller calculates the angle deviation $\Delta\theta$ and corrects the velocity and disturbance estimates. Meanwhile, it updates the angle information $\bar{\theta}$ obtained from the encoder every T_m, a constant or variable sampling cycle. Most of the methods proposed assume that the velocity value is constant during T_m. However, if the acceleration torque $e(t)$ can be estimated as $\hat{e}(m,k)$, the controller can estimate the velocity in a shorter period by integrating the estimate. Therefore, it

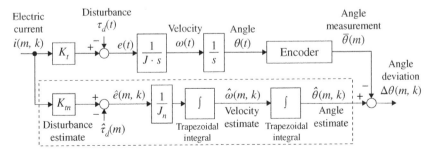

Figure 12.8 Instantaneous speed observer algorithm.

sets a sampling period T shorter than the encoder sampling period T_m and estimates the acceleration torque $\hat{e}(m, k)$ from Equation (12.19):

$$\hat{e}(m, k) = K_{tn}i(m, k) - \hat{\tau}_d(m) \tag{12.19}$$

where (m, k) denotes the kth sample point within the mth encoder reading period, K_{tn} is the nominal value of the torque constant, $i(m, k)$ is the electrical current value and is held for T [s], and $\hat{\tau}_d(m)$ is the disturbance torque assumed constant in the mth encoder reading period. Trapezoidal integration of this yields the velocity estimate, Equation (12.20):

$$\hat{\omega}(m, k) = \hat{\omega}(m, k - 1) + \frac{T}{2}\left\{\frac{\hat{e}(m, k) + \hat{e}(m, k - 1)}{J_n}\right\} \tag{12.20}$$

J_n denotes the nominal value of the moment of inertia. Then, Equation (12.21) expresses the angle estimate between the encoder sampling times:

$$\hat{\theta}(m, k) = \hat{\theta}(m, k - 1) + \frac{T}{2}\{\hat{\omega}(m, k) + \hat{\omega}(m, k - 1)\} \tag{12.21}$$

The deviation $\Delta\theta$ between the estimated and the measured values becomes Equation (12.22) when reading the encoder:

$$\Delta\theta(m, k) = \hat{\theta}(m, T_m/T) - \bar{\theta}(m) \tag{12.22}$$

Assume that $\Delta\theta$ is due to two factors. We interpret it as caused by the error $\Delta\omega$ in the initial velocity at the beginning of the estimation in the interval m and the disturbance estimation error $\Delta\tau_d$. Then, the ratios γ_1 and γ_2 share the responsibility:

$$\gamma_1\Delta\theta = T_m\Delta\omega \tag{12.23}$$

$$\gamma_2\Delta\theta = \frac{T_m^2}{2J_n}\tau_d \tag{12.24}$$

Then, it updates $\Delta\omega$ and $\Delta\tau_d$ with the equations $\Delta\omega = \gamma_1/T_m \cdot \Delta\theta$ and $\Delta\tau_d = 2\gamma_2 J_n/T_m^2 \cdot \Delta\theta$.

In the next interval $m + 1$, it updates $t\hat{a}u(m + 1)$ and $\hat{\omega}(m + 1, 0)$ with the following equations:

$$\hat{T}_d(m + 1) = \hat{T}_d(m) - \Delta T_d \tag{12.25}$$

$$\hat{\omega}(m + 1, 0) = \hat{\omega}(m, K) - \frac{T_m}{J_n}\Delta T_d - \Delta\omega \tag{12.26}$$

Moreover, modify them repeatedly from $t = (m + 1)T_m + 0$ every T[s] using Equations (12.20) and (12.21). The poles of the instantaneous velocity observer can be set based on the liability sharing ratio γ_1, γ_2.[9]

---

9 The instantaneous velocity observer estimation method is equivalent to the multirate disturbance observer (Method 2) described in Section 8.3. Furthermore, reference [7] details the relationship between higher order disturbances and the poles and gamma of the minimal order observer.

References

1 Toshiaki Tsuji, Mariko Mizuochi, Hiroaki Nishi, Kouhei Ohnishi: A Velocity Measurement Method for Acceleration Control, 31st Annual Conference of IEEE Industrial Electronics Society (IECON 2005), 1943–1948, 2005.

2 Toshiaki Tsuji, Takuya Hashimoto, Hiroshi Kobayashi, Mariko Mizuochi, Kouhei Ohnishi: A wide-range velocity measurement method for motion control, IEEE Transactions on Industrial Electronics, Vol. 56, No. 2, 510–519, 2009.

3 Hiroyuki Nagatomi, Kouhei Ohnishi: Acceleration estimation Method for Motion Control System with Optical Encoder, IEEE 2006 International Conference on Industrial Technology, 1480–1485, 2006.

4 Tsutomu Ohmae, Toshihiko Matsuda, Kenzo Kamiyama, Makoto Tachikawa: A microprocessor-controlled high-accuracy wide-range speed regulator for motor drives, IEEE Transactions on Industrial Electronics, Vol. 29, No. 3, 207–211, 1982.

5 Richard C. Kavanagh: Improved digital tachometer with reduced sensitivity to sensor nonideality, IEEE Transactions on Industrial Electronics, Vol. 47, No. 4, 890–897, 2000.

6 Richard Bonert: Design of a high performance digital tachometer with a microcontroller, IEEE Transactions on Industrial Electronics, Vol. 38, No. 6, 1104–1108, 1989.

7 Yoichi Hori: High Performance Control of Servomotors with Low Precision Shaft Encoder using Instantaneous Speed Observer and Adaptive Iidentification of Inertia Moment, Proceedings of 1993 Asis-Pasific Workshop on Advanced Motion Control, 7–12, 1993.

Appendix A

Mathematical Foundations and Control Theory

A.1 Mathematics

A.1.1 Definition and Calculus of Matrix Exponential Functions

The **matrix exponential function** is defined as Equation (A.1):

$$e^{At} = I + At + \frac{1}{2!}A^2t^2 + \frac{1}{3!}A^3t^3 + \cdots + \frac{1}{n!}A^nt^n + \cdots \tag{A.1}$$

The time derivative of a matrix exponential function has the same form as the time derivative of a scalar exponential function if each term on the right-hand side of the defining equation is differentiated and summarized:

$$\begin{aligned}\frac{d}{dt}e^{At} &= A + 2\frac{1}{2!}A^2t + 3\frac{1}{3!}A^3t^2 + \cdots + n\frac{1}{n!}A^nt^{n-1} + \cdots \\ &= A \cdot \left(I + At + \frac{1}{2!}A^2t^2 + \frac{1}{3!}A^3t^3 + \cdots + \frac{1}{n!}A^nt^n + \cdots \right) \\ &= Ae^{At} \end{aligned} \tag{A.2}$$

Conversely, the integral is Equation (A.3):

$$\begin{aligned}\int_0^t e^{At}\, dt &= \left[It + \frac{1}{2!}At^2 + \cdots + \frac{1}{(n+1)!}A^nt^{n+1} + \cdots \right]_0^t \\ &= A^{-1}(e^{At} - I) \end{aligned} \tag{A.3}$$

However, this integral holds if the matrix A is regular. The matrix exponential function is used in modern control theory, and its calculus is often used to introduce digital control.

A.1.2 Positive Definite Matrix

If the symmetric matrix $A \in R^{n \times n}$ has a quadratic form $x^T A x > 0$ for vector $x \in R^n$ and $x \neq 0$, then A is a **positive definite matrix**, and we write $A > 0$.[1]

1 Do not read "A is positive," read "A is positive definite."

Disturbance Observer for Advanced Motion Control with MATLAB/Simulink, First Edition. Akira Shimada.
© 2023 The Institute of Electrical and Electronics Engineers, Inc. Published 2023 by John Wiley & Sons, Inc.
Companion website: www.wiley.com/go/disturbanceobserver

A necessary and sufficient condition for $A > 0$ is that all eigenvalues of A are positive. When $x \neq 0$ and $A \geq 0$, it is called a semipositive definite matrix, and when $A \leq 0$, it is a seminegative definite.

A.1.3 Matrix Rank

The **rank** of a matrix is the maximum number of column vectors (or row vectors) of a given matrix that is linearly independent.[2] Here, we note how to find the rank of a matrix. The $r \times r$-square matrix made by removing some rows and columns from an $m \times n$-matrix is called an r-dimensional submatrix. The $r \leq min(m, n)$ means smaller or equal to whichever of m and n is smaller. Now consider an $m \times n$ matrix A. If the determinants of the $r + 1$-dimensional submatrix of matrix A are all 0, and some of the determinants of the r-dimensional submatrix are not 0, we say the rank of matrix A is r and write $rank(A) = r$.

Example A.1.1 Examine the rank of $A = \begin{bmatrix} 3 & 2 & 1 \\ 6 & 4 & 3 \end{bmatrix}$.

Since A is a 2×3 matrix, the rank r satisfies $r \leq 2$. Then, if we construct the determinant of the two-dimensional submatrices $det \begin{bmatrix} 3 & 2 \\ 6 & 4 \end{bmatrix} = 0$, $det \begin{bmatrix} 3 & 1 \\ 6 & 3 \end{bmatrix} \neq 0$, $det \begin{bmatrix} 2 & 1 \\ 4 & 3 \end{bmatrix} \neq 0$,

which would be some of them are not 0. Hence, the rank is 2. If all of determinants are 0, examine lower one-dimensional lower submatrices, in this example.[3]

A matrix A is said to be row (column) full rank if the rank of the horizontal (vertical) matrix is equal to the number of rows (columns).

A.2 Basic Classical Control Theory

A.2.1 Poles and Zeros

Denote the transfer function $G(s) = N(s)/M(s)$. Both $N(s)$ and $M(s)$ are polynomials of Laplace transformation operator s. $M(s) = s^n \alpha_0 + s^{n-1} \alpha_1 + \cdots + \alpha_n = 0$ is a **characteristic equation**. Its roots $\lambda_1, \lambda_2, \ldots, \lambda_n$ are called poles of $G(s)$, and its roots $N(s)$ are called (transfer) zeros of $G(s)$. For the system represented by the transfer function to be stable, it is necessary and sufficient that the real parts of all roots (=poles) of the characteristic equation are negative. Consider the following control system based on this.

---

2 See technical papers for a strict definition of rank.
3 In MATLAB, we can find it by running "$A = [3, 2, 1; 6, 4, 3]; rank(A)$."

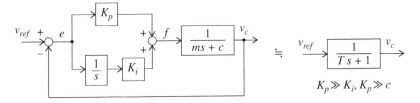

Figure A.1 Block diagram of the PI velocity control system of a cart.

A.2.2 PI Velocity Control

Consider a PI velocity control system such as Figure A.1 for the cart model $P(s) = 1/(ms + c)$. PI control is a control method with a controller using proportional and integral calculations for control error. v_{ref} denotes the velocity reference, and v_c denotes the velocity. The driving force f is the sum of the proportional gain K_p times the error $e = v_{ref} - v_c$ and the error e integrated by $1/s$ and multiplied by the integral gain K_i. By organizing this transfer function, we obtain Equation (A.4).

$$G_c(s) = \frac{v_c(s)}{v_{ref}(s)} = \frac{(Kp + \frac{1}{s}K_i)\frac{1}{ms+c}}{1 + (Kp + \frac{1}{s}K_i)\frac{1}{ms+c}}$$

$$= \frac{K_p s + K_i}{ms^2 + (K_p + c)s + K_i} = \frac{\frac{K_p}{m}s + \frac{K_i}{m}}{s^2 + \frac{K_p+c}{m}s + \frac{K_i}{m}} \tag{A.4}$$

$$= \frac{s + \frac{K_i}{K_p}}{\frac{m}{K_p}s^2 + \frac{K_p+c}{K_p}s + \frac{K_i}{K_p}} \tag{A.5}$$

Let us consider three different design methods.

Method (1) If $K_p \gg K_i$ and $K_p \gg c$, then Equation (A.5) can be approximated by Equation (A.6):

$$G_c(s) = \frac{v_c(s)}{v_{ref}(s)} = \frac{s}{\frac{m}{K_p}s^2 + s} = \frac{1}{\frac{m}{K_p}s + 1} = \frac{1}{Ts + 1} \tag{A.6}$$

In other words, it can be regarded as a first-order delay system. $T = m/K_p$ is a time constant, and K_i can also be determined tunably under the condition $K_p \gg K_i$.[4]

---

4 There are many examples of velocity control system design using such approximations in industry, and they are practical. However, care must be taken because they may not perform well and may diverge when the control plant is complex or subject to fluctuations.

Method (2) This calculates the gain by fitting the characteristic equation to the parameters ζ and ω_n of the second-order delay system. The denominator polynomial of Equation (A.4) is $s^2 + \frac{K_p + c}{m}s + \frac{K_i}{m} = s^2 + 2\zeta\omega_n s + \omega_n^2$, and the gains K_p and K_i are calculated.

Method (3) This uses the characteristic equation, i.e. calculating the gain by specifying the poles of the velocity control system. The denominator polynomial of Equation (A.5) is $s^2 + \frac{K_p + c}{m} + \frac{K_i}{m} = (s - p_1)(s - p_2) = s^2 - (p_1 + p_2)s + p_1 p_2$, and the gains K_p and K_i are calculated. That is, the poles of the velocity control system are P_1 and P_2. Comparing each coefficient, $K_p = -c - m(p_1 + p_2)$ and $K_i = m p_1 p_2$.

A.2.3 PID Position Control System

The PID position control system is a representative example of position control and is shown in Figure A.2. PID control is a control method with a controller using proportional, integral, and derivative calculations for control error. There are several possible design methods for the PID control system design. For example, using the method of specifying the poles as p_1, p_2, and p_3, $K_p = m(p_1 p_2 + p_2 p_3 + p_3 p_1)$, $K_i = -m p_1 p_2 p_3$, and $K_d = -m(p_1 + p_2 + p_3) - c$ are obtained.

A.2.4 Final Value and Initial Value Theorems

The final value theorem is a helpful theorem with which we can calculate the final value of a time function without the response calculation.

> **Theorem A.2.1** (*Final value theorem.*) *The final value of $x(t)$ is calculated as follows:*
>
> $$x(\infty) = \lim_{t \to 0} s \cdot x(s) \tag{A.7}$$
>
> *However, Equation (A.7) can only be applied if the real parts of all poles of $s \cdot x(s)$ are negative. Otherwise, $x(t)$ will not converge, and it diverges to infinity. On the other hand, the initial value theorem is as follows:*
>
> $$x(0+) = \lim_{s \to \infty} s \cdot x(s) \tag{A.8}$$

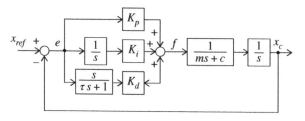

Figure A.2 Block diagram of the PID position control system of a bogie.

The Laplace transform of the time derivative of $x(t)$ is $\int_0^\infty \frac{dx(t)}{dt} e^{-st} dt = s \cdot x(s) - x(0)$. The limit of $s \to 0$ is $\lim_{s \to 0} \int_0^\infty \frac{dx(t)}{dt} e^{-st} dt = \lim_{s \to 0} s \cdot x(s) - x(0)$. Integrating with s on the left side as 0, we get $x(\infty) - x(0) = \lim_{s \to 0} s \cdot x(s) - x(0)$. If we omit $x(0)$ on the left and right, we have Equation (A.7).[5] The proof of the initial value theorem is almost the same, only the limit should be ∞ instead of 0.

A.3 Basic Modern Control Theory

This section summarizes the minimum necessary to understand this book on the fundamentals of modern control theory and observer design [1].

A.3.1 State and Output Equations

Generally, assume state variables $x(t) \in R^n$, control input $u(t) \in R^m$, and observed output $y(t) \in R^l$. Equation (A.9) represents the equation of state of a linear time-invariant system, and the output equation is Equation (A.10):

$$\dot{x}(t) = Ax(t) + Bu(t) \tag{A.9}$$

$$y(t) = Cx(t) + Du(t) \tag{A.10}$$

However, each coefficient matrix is $A \in R^{n \times n}$, $B \in R^{n \times m}$, $C \in R^{l \times n}$, and $D \in R^{l \times m}$.

In this case, the transfer function is obtained by $G(s) = C(sI - A)^{-1}B + D$, but notation $G(s) = [A, B, C, D]$ or $G(s) = [A, B, C, D]$ or $G(s) = \begin{bmatrix} A & B \\ \hline C & D \end{bmatrix}$ is adequately selected depending on the situation. The latter is specifically referred to as Doyle's notation.

$D = 0$ is often the case in many mechatronic systems, and $+Du(t)$ is omitted in this book, except in Chapter 3 and Section A.4.

Example A.3.1 (State and output equations of the cart model with position observation.) Many mechatronic systems are often equipped with position sensors. Let us take the cart model as an example to create the state and output equations.

Assume state variable $x(t) = [x_c(t), v_c(t)]^T$, input $u(t) = f(t)$, and output $y(t) = x_c(t)$. Since $\dot{x} = [\dot{x}_c(t), \dot{v}_c(t)]^T = [v_c, a_c]^T$, leaving the first component $\dot{x}_c(t) = v_c(t)$ intact, we obtain $\dot{v}_c(t) = a_c(t) = 1/m \cdot f(t)$. Then, organizing them in a matrix form, the state and output equations become Equations (A.11) and (A.12):

$$\dot{x}(t) = \begin{bmatrix} \dot{x}_c(t) \\ \dot{v}_c(t) \end{bmatrix} = Ax(t) + Bu(t)$$

$$= \begin{bmatrix} 0 & 1 \\ 0 & 0 \end{bmatrix} \begin{bmatrix} x_c(t) \\ v_c(t) \end{bmatrix} + \begin{bmatrix} 0 \\ 1/m \end{bmatrix} f(t) \tag{A.11}$$

$$y(t) = Cx(t) = \begin{bmatrix} 1 & 0 \end{bmatrix} \begin{bmatrix} x_c(t) \\ v_c(t) \end{bmatrix} \tag{A.12}$$

---

5 Strictly speaking, in Equation (A.8), the $t = 0+$ s represents the time immediately after the start.

A.3.2 Solution of the State Equation for the Continuous System

The state equation $\dot{x}(t) = Ax(t) + Bu(t)$ is a linear differential equation, and it is rare when we should find the solution directly, but here, we can find the general solution by assuming that the solution form is Equation (A.13):

$$x(t) = e^{At}\{x(0) + z(t)\} \in R^n \qquad (A.13)$$

Differentiating both sides by time gives $\dot{x} = Ae^{At}\{x(0) + z(t)\} + e^{At}\dot{z}(t) = Ax(t) + e^{At}\dot{z}(t)$. Therefore, focusing on the second term on the right, $Bu(t) = e^{At}\dot{z}(t)$. That is, integrating $\dot{z}(t) = e^{-At}Bu(t)$, we obtain $z(t) = \int_0^t e^{-A\tau}Bu(\tau)d\tau$. Substituting this into Equation (A.13), we obtain the following solution:

$$x(t) = e^{At}x(0) + \int_0^t e^{A(t-\tau)}Bu(\tau)d\tau \qquad (A.14)$$

The output is $y(t) = Cx(t) = Ce^{At}x(0) + C\int_0^t e^{A(t-\tau)}Bu(\tau)d\tau$.

A.3.3 Equation of State to Transfer Function

Let us transform the state equation $\dot{x}(t) = Ax(t) + Bu(t)$ by Laplace transform. However, assuming the initial value $x(0) = 0$, $sx(s) - x(0) = sx(s) = Ax(s) + Bu(s)$. That is, $x(s) = (sI - A)^{-1}Bu(s)$ via $(sI - A)x(s) = Bu(s)$. If $y(t) = Cx(t)$ is also Laplace transformed to $y(s) = Cx(s)$, we obtain Equation (A.15):

$$y(s) = C(sI - A)^{-1}Bu(s) \qquad (A.15)$$

A.3.4 Poles and Zeros of Continuous Systems

The definitions of poles and zeros are as follows:

Definition A.3.1 The roots $\lambda_1, \lambda_2, \dots, \lambda_n$ of $det(sI - A) = 0$ are called poles (pole). If the real parts of all poles are negative (positive), the system is stable (unstable). Pole with a negative real part is called a stable pole, and a positive pole is called an unstable pole.

Definition A.3.2 In the system matrix $P(s) = \begin{bmatrix} A - sI & B \\ C & 0 \end{bmatrix}$, the A point $s = s_0$ satisfying $rank[P(s_0)] < min(n + m, n + l)$ is called an (invariant) zero. Zero with a negative real part is called a stable zero, and a positive zero is called an unstable zero.

When the system satisfies a property called controllability, introduced in the next Section A.3.5., the poles can be set arbitrarily in feedback control. However, the zeros are not and are therefore called invariant zeros.

A.3.5 Controllability and Observability of Continuous Systems

Definitions of controllable and observable are as follows:

Definition A.3.3   Suppose that the initial value $x(0)$ of the state variable $x(t)$ and the final value x_e are arbitrarily determined. If $x(t_e) = x_e$ can be achieved with some control input $u(t)$ and finite time $t_e(> 0)$, we say the system (A, B) is controllable.

Theorem A.3.1   *The fact that the system is controllable is equivalent to the following conditions being satisfied:*
(1) The rank of the controllability matrix $U_c = [B, AB, A^2B, \ldots, A^{n-1}B]$ is n.
(2) There exists a gain matrix F with arbitrary eigenvalues of $A - BF$.

If the system is controllable, the state feedback $u(t) = -Fx(t)$ can set all poles arbitrarily and stabilize the system.

Definition A.3.4   If the initial value $x(0)$ of the state variable is uniquely determined from the results of the control input $u(t)$ and the observed output $y(t)$ from $t = 0$ to time $t_e(> 0)$, we say the system (A, C) is observable.

Theorem A.3.2   *The fact that the system is observable is equivalent to the following condition:*

(1) The rank of the observability matrix $U_0 = \begin{bmatrix} C \\ CA \\ CA^2 \\ \vdots \\ CA^{n-1} \end{bmatrix}$ is n.

(2) There exists a matrix H with arbitrary eigenvalues of $A - HC$.

If the system is observable, an observer can be designed to estimate the state variable $x(t)$.

A.3.6 Duality Theorem

The duality theorem is also important.

> **Theorem A.3.3** *The equations "(A,B) is controllable" and "(B^T, A^T) is observable" are equivalent, and "(C, A) is observable" and "(A^T, C^T) is controllable" are equivalent.*

Given that the rank is the same after transposing the matrix, we obtain the transposed matrix of the controllability matrix U_c:

$$U_c^T = \begin{bmatrix} B & AB & A^2B & \cdots & A^{n-1}B \end{bmatrix}^T = \begin{bmatrix} B^T \\ B^T A^T \\ \vdots \\ B^T (A^{n-1})^T \end{bmatrix} = \begin{bmatrix} B^T \\ B^T A^T \\ \vdots \\ B^T (A^T)^{n-1} \end{bmatrix}.$$

The second equation on the right-hand side is the (B^T, A^T) observable matrix, whose rank is equal to $rank(U_c)$, so if (A, B) controllable, then (B^T, A^T) is observable. The same is true for the latter. This duality can be used in observer design.

A.3.7 State Feedback Control of Continuous Systems

State Feedback Control and Pole Assignment Method

The feedback function that detects the state variable $x(t)$ and multiplies it by the gain matrix F to make the control input $u(t) = -Fx(t)$ is called state feedback. The block diagram is shown in Figure A.3. $\dot{x}(t) = Ax(t) + Bu(t) = Ax(t) + B\{-Fx(t)\} = (A - BF)x(t)$. To stabilize the control system, the feedback gain matrix F is determined so that the real parts of all eigenvalues of $A - BF$, that is the real parts of the poles of the control system, are negative values. This method is called the **pole assignment method**.[6]

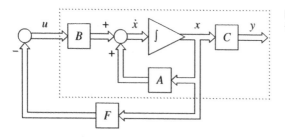

Figure A.3 Structure of state feedback control system.

---

6 The pole assignment method is also called the **pole placement method**.

Example A.3.2 For the cart represented by Equation (A.11), the driving force $f(t)$ is considered to be the control input $u(t)$, and $f = -[F_1, F_2] \cdot [x_c, v_c]^T$ is applied:

$$
\begin{aligned}
\begin{bmatrix} \dot{x}_c(t) \\ \dot{v}_c(t) \end{bmatrix} &= \begin{bmatrix} 0 & 1 \\ 0 & 0 \end{bmatrix} \begin{bmatrix} x_c(t) \\ v_c(t) \end{bmatrix} + \begin{bmatrix} 0 \\ 1/m \end{bmatrix} f(t) \\
&= \begin{bmatrix} 0 & 1 \\ 0 & 0 \end{bmatrix} \begin{bmatrix} x_c(t) \\ v_c(t) \end{bmatrix} - \begin{bmatrix} 0 \\ 1/m \end{bmatrix} [F_1 \ F_2] \begin{bmatrix} x_c(t) \\ v_c(t) \end{bmatrix} \\
&= \begin{bmatrix} 0 & 1 \\ -\frac{F_1}{m} & -\frac{F_2}{m} \end{bmatrix} \begin{bmatrix} x_c(t) \\ v_c(t) \end{bmatrix}
\end{aligned}
\tag{A.16}
$$

Calculate the poles from the characteristic polynomial (=denominator polynomial):

$$
\{sI - (A - BF)\}^{-1} = \begin{bmatrix} s & -1 \\ \frac{F_1}{m} & s + \frac{F_2}{m} \end{bmatrix}^{-1} = \frac{\begin{bmatrix} s + \frac{F_1}{m} & 1 \\ -\frac{F_1}{m} & s \end{bmatrix}}{s^2 + s\frac{F_2}{m} + \frac{F_1}{m}}
$$

For the control system to be stable, the real parts of all poles must be negative. Replace $\frac{F_2}{m} = -(p_1 + p_2)$ and $\frac{F_1}{m} = p_1 p_2$, and $F_1 = mp_1 p_2$ and $F_2 = -m(p_1 + p_2)$ can be obtained if p_1 and p_2 are given by the designer.[7]

Optimal Control

Stable feedback gains can be obtained relatively freely using the pole assignment method. However, it is quite difficult to decide where to place the poles adequately. Therefore, a control method that determines an appropriate evaluation function and finds the input that minimizes the evaluation function has been proposed. This is called optimal or LQ control because it uses a Linear Quadratic evaluation function.

Definition A.3.5 (Optimal regulator.) The control input that minimizes the evaluation function of Equation (A.17) and still stabilizes the control system is given by Equations (A.18) and (A.19). The variable P is the solution of Equation (A.20), called the Riccati equation, which is a positive definite and

(continued)

---

7 In MATLAB, the poles of the control system are, for example, $pole = [p_1, p_2] = [-2, -3]$, and the control gain can be obtained with $F = place(A, B, pole)$. The $place$ function is based on the pole cart assignment method. However, a $place$ does not accept multiple poles, so different poles must be specified.

(*Continued*)

symmetric matrix. Choose Q and R to be semidefinite and positive definite with weight matrices for the state variables and control inputs.

$$J = \int_0^\infty x^T(t)Qx(t) + u^T(t)Ru(t)dt \tag{A.17}$$

$$u(t) = -Fx(t) \tag{A.18}$$

$$F = R^{-1}B^T P \tag{A.19}$$

$$A^T P + PA - PBR^{-1}B^T P + Q = 0 \tag{A.20}$$

Equation (A.20) has the only solution when (A, B) is controllable, and the minimum value of Equation (A.17) is $J = x^T(0)Px(0)$.

Q is a diagonal matrix with all elements positive. From top left to bottom right, weights are expressed in the order of state variables, with larger weights for state variables that converge faster. The ratio of the weights should be about 10 times the factorial of the state variable.[8]

Using a cart as an example, consider the meaning of optimal control. Since $x(t) = [x_c(t), v_c(t)]^T$ and $u(t) = f(t)$, if $Q = diag(Q_1, Q_2)$, then $x^T(t)Qx(t) + u^T(t)Ru(t) = \begin{bmatrix} x_c(t) & v_c(t) \end{bmatrix} \begin{bmatrix} Q_1 & 0 \\ 0 & Q_2 \end{bmatrix} \begin{bmatrix} x_c(t) \\ v_c(t) \end{bmatrix} + Rf(t)^2 = Q_1 x_c(t)^2 + Q_2 v_c(t)^2 + Rf(t)^2$.

If its integral value is finite and can be minimized, the absolute value of $x_c(t)$ is smaller when Q_1 is large, that of $v_c(t)$ is smaller when Q_2 is large, and that of $u(t)$ is smaller when R is large.

Identity Observer

To perform state feedback, the sensor must observe all of the elements in $x(t)$, but what can actually be observed is the observed output $y(t)$. For $y(t) = Cx(t) = x(t)$, C must be an identity matrix, but this is rare. So, we need to consider a method to estimate $x(t)$. State estimation is possible only if the system is observable, and the estimation function is called an observer. Specifically, identity observers, which have the same dimension as the control plant expressed as Equations (A.9) and (A.10), are represented by Equations (A.21) and (A.22):

$$\dot{\hat{x}}(t) = A\hat{x}(t) + Bu(t) + H\{y(t) - \hat{y}(t)\} \tag{A.21}$$

$$\hat{y}(t) = C\hat{x}(t) \tag{A.22}$$

---

8 To design the optimal regulator in MATLAB, for example, $A = 0, 1; -2, -3]; B = 0; 1]; Q = diag$ $([10, 2]); R = 1; F = lqr(A, B, Q, R)$, and the optimal control gain $F = [1.74, 0.81]$.

Figure A.4 Block diagram of control system with identity observer.

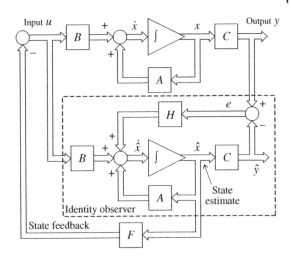

where the state estimate $\hat{x}(t) \in R^n$, the output estimate $\hat{y}(t) \in R^l$, and the observer gain $H \in R^{n \times l}$. Let the estimation error of the real system and the observer $e(t) = x(t) - \hat{x}(t)$, and the time derivative $\dot{e} = \dot{x} - \dot{\hat{x}} = (Ax + Bu) - (A\hat{x} + Bu + H\{y - \hat{y}\}) = A(x - \hat{x}) + HC(x - \hat{x}) = (A - HC)e$ is obtained. The eigenvalues of $A - HC$ are called the poles of the observer, and if the real parts of all the poles are negative, then over time $e(t) = exp\{(A - HC)t\}e(0)$ converges to 0. That is, $\hat{x}(t)$ converges to $x(t)$.[9] Figure A.4 shows the block diagram of the control system using both the identity observer and state feedback.

Minimal Order Observer

The basic theory of minimal order observers and the Gopinath design method are presented. Suppose that the state and output equations for an m-input l-output n-order system are represented by Equations (A.23) and (A.24):

$$\dot{x}(t) = Ax(t) + Bu(t) \tag{A.23}$$

$$y(t) = Cx(t) \tag{A.24}$$

where $A \in R^{n \times n}$, $B \in R^{m \times n}$, and $C \in R^{l \times n}$.

If l outputs correspond to l state variables and can be observed unaffected by noise, there is no need to estimate l outputs. The remaining $n - l$ state variables need only be estimated, and an observer with the smallest dimension can be designed.

---

9 $F = place(A, B, pole)$ is the control gain using the place function of MATLAB. For example, by placing pole_ob=[-5, -6] for the desired observer pole and getting the temporary gain $H_{temp} = place(A', C', pole_o b)$, we can get the observer gain from $H = H_t emp'$ considering the duality.

[Step 1]

Choose an $(n - l) \times n$ matrix U such that $S = \begin{bmatrix} C \\ U \end{bmatrix}$, $det(S) \neq 0$, and assume a new state variable $z(t) \in R^{n-l}$, which yields Equation (A.25):

$$\begin{bmatrix} y(t) \\ z(t) \end{bmatrix} = Sx(t) = \begin{bmatrix} C \\ U \end{bmatrix} x(t) \tag{A.25}$$

From Equation (A.25), Equation (A.26) holds, but $z(t)$ has no physical meaning.

$$x(t) = S^{-1} [y(t)] = \begin{bmatrix} C \\ U \end{bmatrix}^{-1} \begin{bmatrix} y(t) \\ z(t) \end{bmatrix} \tag{A.26}$$

[Step 2]

Assume a new equation of the state equation (A.27), where the coefficient matrix $\hat{A} \in R^{(n-l) \times (n-l)}$, $\hat{B} \in R^{(n-l) \times l}$, and $\hat{J} \in R^{(n-l) \times m}$:

$$\dot{z}(t) = \hat{A}z(t) + \hat{B}y(t) + \hat{J}u(t) \tag{A.27}$$

From Equations (A.23), (A.24), and (A.27), we obtain Equation (A.28):

$$\begin{aligned}\dot{z} - U\dot{x} &= \hat{A}z + \hat{B}y + \hat{J}u - U(Ax + Bu) \\ &= \hat{A}(z - Ux) + (\hat{A}U + \hat{B}C - UA)x + (\hat{J} - UB)u \end{aligned} \tag{A.28}$$

If we choose Equations (A.29) and (A.30) to hold, then $\dot{z}(t) - U\dot{x}(t) = \hat{A}(z(t) - Ux(t))$; furthermore, through $z(t) - Ux(t) = e^{\hat{A}t}(z(0) - Ux(0))$, we obtain Equation (A.31).

$$\hat{A}U + \hat{B}C = UA \tag{A.29}$$

$$\hat{J} = UB \tag{A.30}$$

$$z(t) = Ux(t) + e^{\hat{A}t}(z(0) - Ux(0)) \tag{A.31}$$

If we choose $\hat{A}$ with stable eigenvalues, then $z(t) \rightarrow Ux(t)$ $(t \rightarrow \infty)$.

[Step 3]

Assume that $\hat{C}$ and $\hat{D}$ satisfy Equations (A.32) and (A.33).

$$\hat{C}U + \hat{D}C = I_n \tag{A.32}$$

$$\hat{x}(t) = \hat{C}z(t) + \hat{D}y(t) \tag{A.33}$$

If Equation (A.32) and $z(t) \rightarrow Ux(t)$ $(t \rightarrow \infty)$ hold, $\hat{C}z(t) + \hat{D}y(t) \rightarrow \hat{C}Ux(t) + hatDy(t) = (I_n - \hat{D}C)x(t) + \hat{D}Cx(t) = x(t)$.

Consequently, it can be observed that Equation (A.33) is an expression for estimating $x(t)$. Illustrating the above, we obtain Figure A.5.

Equations (A.29), (A.30), and (A.32) can be summarized in $\begin{bmatrix} \hat{A} & \hat{B} \\ \hat{C} & \hat{D} \end{bmatrix} \begin{bmatrix} U \\ C \end{bmatrix} = \begin{bmatrix} UA \\ I_n \end{bmatrix}$ or $\begin{bmatrix} \hat{A} & \hat{B} \\ \hat{C} & \hat{D} \end{bmatrix} = \begin{bmatrix} UA \\ I_n \end{bmatrix} \begin{bmatrix} U \\ C \end{bmatrix}^{-1}$.

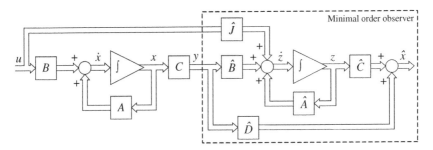

Figure A.5 Block diagram of the continuous system minimal order observer.

Gopinath Design Methods

We introduce the typical design method of Gopinath [2].

[Step 1]

First, coordinate transformations are performed in preparation for the minimal order observer design. Choose an $(n - l) \times n$ matrix W with $T = \begin{bmatrix} C \\ W \end{bmatrix}$, $det(T) \neq 0$,

and $w(t) = \begin{bmatrix} y(t) \\ z(t) \end{bmatrix} = \begin{bmatrix} C \\ W \end{bmatrix} x(t)$. Similarly, $x = T^{-1}w$ with.[10] Multiply both sides of

$\dot{x} = Ax + Bu$ by T from the left side, let $\dot{w} = TAT^{-1}w + TBu = \bar{A}w + \bar{B}u$, and let $y = Cx = CT^{-1}w = \bar{C}w$, then the equations of state and output equations after a coordinate transformation yield Equations (A.34) and (A.35):

$$\dot{w} = \begin{bmatrix} \dot{y} \\ \dot{z} \end{bmatrix} = \bar{A}w + \bar{B}u = \begin{bmatrix} A_{11} & A_{12} \\ A_{21} & A_{22} \end{bmatrix} \begin{bmatrix} y \\ z \end{bmatrix} + \begin{bmatrix} B_1 \\ B_2 \end{bmatrix} u \tag{A.34}$$

$$y = \bar{C}w \tag{A.35}$$

where $A_{11} \in R^{l \times l}$, $A_{12} \in R^{l \times (n-l)}$, $A_{21} \in R^{(n-l) \times l}$, and $A_{22} \in R^{(n-l) \times (n-l)}$. Also, since $\bar{C} = CT^{-1}$, we have $\bar{C} = CT^{-1}$, and $C = \bar{C}T = \begin{bmatrix} * & * \end{bmatrix} \begin{bmatrix} C \\ W \end{bmatrix}$ to $\bar{C} = \begin{bmatrix} I_l & 0_{l \times (n-l)} \end{bmatrix}$ must hold. Decomposing Equation (A.34) yields Equations (A.36)–(A.38), which we call Gopinath form:

$$\dot{y} = A_{11}y + A_{12}z + B_1u \tag{A.36}$$

$$\dot{z} = A_{21}y + A_{22}z + B_2u \tag{A.37}$$

$$y = \bar{C}w = \begin{bmatrix} I_l & 0_{l \times (n-l)} \end{bmatrix} \cdot w \tag{A.38}$$

---

10 Here $T = S$, i.e., we could have chosen the same character U instead of W, the U was used in Step 1 in the subsection Minimal order observer. However, we denied them separately to avoid confusion with the definitions and computational process in the next step.

[Step 2]

Now, we design a minimal order observer for the coordinate transformed equations (A.34) and (A.35). Assume a new matrix $\bar{U}$ satisfying $rank \begin{bmatrix} C \\ \bar{U} \end{bmatrix} = n$ as in Equation (A.39):

$$\bar{U} = \begin{bmatrix} -L_{(n-l)\times l} & I_{n-l} \end{bmatrix} \tag{A.39}$$

Corresponding to Equation (A.29), we obtain Equation (A.40), which yields Equation (A.41) when decomposed:

$$\hat{A}\bar{U} + \hat{B}\overline{C} = \hat{A} \begin{bmatrix} -L & I_{n-l} \end{bmatrix} + \hat{B} \begin{bmatrix} I_l & 0_{l\times(n-l)} \end{bmatrix}$$

$$= \bar{U}\bar{A} = \begin{bmatrix} -L & I_{n-l} \end{bmatrix} \begin{bmatrix} A_{11} & A_{12} \\ A_{21} & A_{22} \end{bmatrix} \tag{A.40}$$

$$\begin{bmatrix} -\hat{A}L + \hat{B} & \hat{A} \end{bmatrix} = \begin{bmatrix} -LA_{11} + A_{21} & -LA_{12} + A_{22} \end{bmatrix} \tag{A.41}$$

We obtain Equations (A.42) and (A.43) by comparing both sides:

$$\hat{A} = A_{22} - LA_{12} \tag{A.42}$$

$$\hat{B} = A_{21} - LA_{12} + \hat{A}L \tag{A.43}$$

Similarly, we obtain Equation (A.44), corresponding to Equation (A.30):

$$\hat{J} = \bar{U}\overline{B} = \begin{bmatrix} -L & I_{n-l} \end{bmatrix} \begin{bmatrix} B_1 \\ B_2 \end{bmatrix} = -LB_1 + B_2 \tag{A.44}$$

[Step3]

The formula equation (A.32) is applied, and the subformula is obtained:

$$\hat{C}\bar{U} + \hat{D}\overline{C} = \hat{C} \begin{bmatrix} -L & I_{n-l} \end{bmatrix} + \hat{D} \begin{bmatrix} I_l & 0_{l\times(n-l)} \end{bmatrix}$$

$$= \begin{bmatrix} -\hat{C}L + \hat{D} & \vdots & \hat{C} \end{bmatrix} = I_n = \begin{bmatrix} I_l & 0_{l\times(n-l)} \\ 0_{(n-l\times l)} & I_{n-l} \end{bmatrix}$$

From this, $\hat{C} = \begin{bmatrix} 0l \times (n-l) \\ I_{n-l} \end{bmatrix}$, and $\hat{D} = \begin{bmatrix} I_l \\ 0_{(n-l)\times l} \end{bmatrix} + \hat{C}L = \begin{bmatrix} I_l \\ L_{(n-l)\times l} \end{bmatrix}$ is determined, and we obtain the following equation corresponding to Equation (A.33): $\hat{w} = \hat{C}z + \hat{D}y$. However, finally, noting that in this section we have designed a minimal order observer that estimates $w(t)$ rather than a minimal order observer that estimates $x(t)$, we obtain the relation $X = T^{-1}w$, $\hat{X} = T^{-1}\hat{C}z + T^{-1}\hat{D}y$.

Once again, replacing $T^{-1}\hat{C}$ with $\hat{C}$ and $T^{-1}\hat{D}$ with $\hat{D}$, we obtain $\hat{C} = T^{-1}\begin{bmatrix} 0_{l\times(n-l)} \\ I_{n-l} \end{bmatrix}$. If $\hat{D} = T^{-1}\begin{bmatrix} I_l \times l_{(n-l)\times l} \end{bmatrix}$, then we obtain an estimating equation (A.45) for the state variable x:

$$\hat{x}(t) = \hat{C}z(t) + \hat{D}y(t) \tag{A.45}$$

If we apply this to Figure A.5, we can construct the minimal order observer.[11]

A.3.8 Servo System Design

A servo system is a control system in which the error $e(t) = r(t) - y(t)$ between the reference value $r(t)$ and the observed output $y(t)$ converges to 0. Basically, the integral value w of the deviation e is assumed, and an extended system is created using the extended state variable $\tilde{x}(t) = [x(t), w(t)]^T$.

The Design of Type 1 Service System by Sequence

$$\dot{\tilde{x}}(t) = \tilde{A}\tilde{x}(t) + \tilde{B}u(t) - \tilde{B}d(t) + \tilde{R}r(t) \tag{A.46}$$

where $\tilde{x}(t) = \begin{bmatrix} x(t) \\ \hline w(t) \end{bmatrix}$, $\tilde{A} = \begin{bmatrix} A & 0 \\ \hline -C & 0 \end{bmatrix}$, $\tilde{B} = \begin{bmatrix} B \\ \hline 0 \end{bmatrix}$, and $\tilde{R} = \begin{bmatrix} 0 \\ \hline I \end{bmatrix}$. Define the following equation at $t = \infty$:

$$\dot{\tilde{x}}(\infty) = \tilde{A}\tilde{x}(\infty) + \tilde{B}u(\infty) - \tilde{B}d(\infty) + \tilde{R}r(\infty) \tag{A.47}$$

If $\dot{d}(t) = 0$, $\dot{r}(t) = 0$, and subtract both sides of Equations (A.46) and (A.47) from each other, then $d(t) - d(\infty) = 0, r(t) - r(\infty) = 0$, and $\tilde{x}_e = [x(t) - x(\infty)w(t) - w(\infty)]^T$.

Letting $u_e = u(t) - u(\infty)$, we obtain the extended deviation system equation (A.48):

$$\dot{\tilde{x}}_e = \tilde{A}\tilde{x}_e + \tilde{B}u_e \tag{A.48}$$

If the system is controllable, the state feedback $u_e = -\tilde{F}\tilde{x}_e$ provides a gain matrix $\tilde{F} = [F_1, F_2]$ that stabilizes the extended system error system.

---

11 From the design where the real parts of the eigenvalues of $\hat{A} = A_{22} - LA_{12}$ are negative, we can see that the process of designing $A - HC$ is recalled. Just as H is the gain of the same-dimensional observer, L is the gain of estimating $z(t)$ for the minimal order observer.

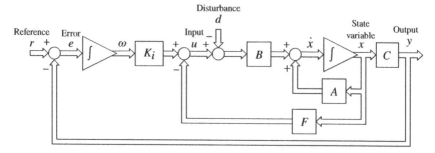

Figure A.6 Block diagram of continuous servo system.

$u_e = u(t) - u(\infty) = -[F_1, F_2] \begin{bmatrix} x(t) - x(\infty) \\ w(t) - w(\infty) \end{bmatrix}$, and the control input is $u(t) = -F_1x(t) - F_2w(t) + \{u(\infty) + F_1x(\infty) + F_2w(\infty)\}$. If $u(\infty) + F_1x(\infty) + F_2w(\infty) = 0$, the control input is $u(t) = -Fx(t) + K_iw(t)$. The block diagram of the basic servo system is shown in Figure A.6.

A.4 Doyle's Notation and Double Coprime Factorization

A.4.1 Doyle's Notation

This section introduces the formulas necessary to understand Chapter 3 [3].

Expressing $G(s) = \begin{bmatrix} A & B & C & D \end{bmatrix}$ using **Doyle's notation** $G(s) = \begin{bmatrix} A & B \\ \hline C & D \end{bmatrix}$, we have

(1) Sign exchange $\begin{bmatrix} A & B \\ \hline -C & D \end{bmatrix} = \begin{bmatrix} A & -B \\ \hline C & D \end{bmatrix}$

(2) Similarity transformation $\begin{bmatrix} T^{-1}AT & T^{-1}B \\ \hline CT & D \end{bmatrix} = \begin{bmatrix} A & B \\ \hline C & D \end{bmatrix}$

(3) Series connection $\begin{bmatrix} A_1 & B_1 \\ \hline C_1 & D_1 \end{bmatrix} \times \begin{bmatrix} A_2 & B_2 \\ \hline C_2 & D_2 \end{bmatrix} = \begin{bmatrix} A_1 & B_1C_2 & B_1D_2 \\ 0 & A_2 & B_2 \\ \hline C_1 & D_1C_2 & D_1D_2 \end{bmatrix}$

(4) Delete uncontrollable subspace (if A_2 is stable)

$\begin{bmatrix} A_11 & A_{12} & B_1 \\ 0 & A_2 & 0 \\ \hline C_1 & C_2 & D \end{bmatrix} = \begin{bmatrix} A_1 & B_1 \\ \hline C_1 & D \end{bmatrix}$

(5) Delete unobservable subspace (if A_2 is stable)

$\begin{bmatrix} A_1 & 0 & B_1 \\ 0 & A_2 & B_2 \\ \hline C_1 & 0 & D \end{bmatrix} = \begin{bmatrix} A_1 & B_1 \\ \hline C_1 & D \end{bmatrix}$

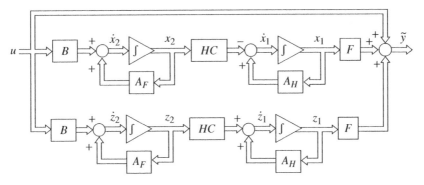

Figure A.7 Block diagram of $\tilde{X}D + \tilde{Y}N = I$.

A.4.2 Confirmation of Double Coprime Factorization

Applying Doyle's notation to check Equation (3.19), we obtain

$$\tilde{X}D + \tilde{Y}N = \begin{bmatrix} A_H & -HC & 0 \\ 0 & A_F & B \\ \hline F & 0 & I \end{bmatrix} + \begin{bmatrix} A_H & HC & 0 \\ 0 & A_F & B \\ \hline F & 0 & 0 \end{bmatrix} \quad (A.49)$$

Decoding the expression (A.49), the state variable of $\tilde{X}D + \tilde{Y}N$ is tentatively $x = [x_1^T, x_2^T]^T$. Furthermore, the state variable of $\tilde{Y}N$ is $z = [z_1^T, z_2^T]^T$, which is illustrated as Figure A.7.

All cancel each other out, leaving only $\tilde{y} = u$. That is, $\tilde{X}D + \tilde{Y}N = I$.

Similarly,

$$\tilde{N}D - \tilde{D}N = \begin{bmatrix} A_H & -BF & B \\ 0 & A_F & B \\ \hline C & 0 & 0 \end{bmatrix} - \begin{bmatrix} A_H & -BF & B \\ 0 & A_F & B \\ \hline C & 0 & 0 \end{bmatrix} = 0$$

yields $\tilde{N}D - \tilde{D}N = 0$. In the same way, we obtain

$$\tilde{X}Y - \tilde{Y}X = \begin{bmatrix} A_H & HC & H \\ 0 & A_F & H \\ \hline F & 0 & 0 \end{bmatrix} - \begin{bmatrix} A_H & HC & H \\ 0 & A_F & H \\ \hline F & 0 & 0 \end{bmatrix} = 0$$

Finally, as in the confirmation process for Equation (A.49),

$$\tilde{N}Y + \tilde{D}X = \begin{bmatrix} A_H & BF & 0 \\ 0 & A_F & H \\ \hline C & 0 & 0 \end{bmatrix} + \begin{bmatrix} A_H & BF & 0 \\ 0 & A_F & H \\ \hline -C & 0 & 1 \end{bmatrix} = I$$

That is, the relation $\tilde{N}Y + \tilde{D}X = I$ can be confirmed.

A.5 Foundations of Digital Control Theory

A.5.1 Digital Control and State and Output Equations

We introduce a minimal commentary on **digital control theory** [4] necessary to understand this book. The dotted line in Figure A.8 represents the continuous

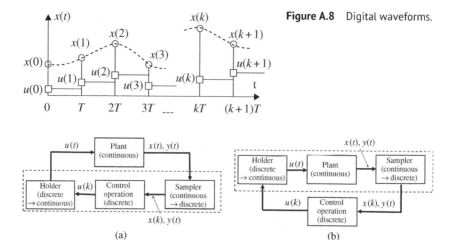

Figure A.8 Digital waveforms.

Figure A.9 Structure of digital control system. (a) Actual structure and (b) interpreted structure.

signal $x(t)$, and the ○ represents the sampled value $x(k)$ at $t = kT$. Meanwhile, the basic structure of the digital control system is described, as in Figure A.9a. The control plant operates continuously, and the controller samples the control plant's continuous output every fixed period (=control cycle) T seconds and performs control operations using the sampled values. The hold function holds the calculated control input $u(t)$ until the next sample time. The waveform of $u(k)$ of Figure A.8 illustrates the hold. Figure A.9 (a) describes the block diagram of the digital control system. When designing the control system, the hold and sampler functions belong to the control plant side as shown in Figure A.9 (b), and the three together are interpreted as a discretized control plant. The inlet and outlet of the dotted line in Figure A.9b have discretized values entering and exiting. Let T be the control period T and $x(t)$ at $t = kT$ be abbreviated as $x(k)$ and $u(t)$ as $u(k)$. Moreover, assume that $u(t)$ is maintained at a constant value $u(k)$ at $kT \leq t < (k + 1)T$. Using Equation (A.14), we obtain Equation (A.50):

$$x(k) = e^{AkT}x(0) + \int_0^{kT} e^{A(kT-\tau)}Bu(\tau)d\tau \tag{A.50}$$

$$x(k+1) = e^{A(k+1)T}x(0) + \int_0^{(k+1)T} e^{\{A(k+1)T-\tau)\}}Bu(\tau)d\tau$$

$$= e^{AT}e^{AkT}x(0) + e^{AT}\int_0^{kT} e^{A(kT-\tau)}Bu(\tau)d\tau$$

$$+ e^{AT}\int_{kT}^{(k+1)T} e^{A(kT-\tau)}Bu(\tau)d\tau$$

$$= e^{AT}\left\{ e^{AkT}x(0) + \int_0^{kT} e^{A(kT-\tau)}Bu(\tau)d\tau \right\}$$

$$+ e^{AT} \int_{kT}^{(k+1)T} e^{A(kT-\tau)} Bu(\tau) d\tau$$

$$= e^{AT} x(k) + \int_{kT}^{(k+1)T} e^{\{A(k+1)T-\tau\}} d\tau Bu(k) \tag{A.51}$$

Replacing $\sigma = (k+1)T$, the substitution integral yields $\tau : kT \to (k+1)T$. Using $\sigma : T \to 0$ and $d\sigma = -d\tau$, we obtain

$$x(k+1) = e^{AT} x(k) + \int_0^T e^{A\sigma} d\sigma Bu(k) \tag{A.52}$$

Finally, replacing $A_d = e^{AT}$, $B_d = \int_0^T e^{A\sigma} d\sigma B$, and $C_d = C$, we obtain the state equation (A.53) and output equation (A.54) for the digital system:[12]

$$x(k+1) = A_d x(k) + B_d u(k) \tag{A.53}$$

$$y(k) = C_d x(k) \tag{A.54}$$

Equations (A.53) and (A.54) are the exact expressions corresponding to the dashed line part of Figure A.9b.

A.5.2 Poles and Zeros of Digital Systems

Define poles and zeros as in a continuous system.

Definition A.5.1   The roots $\lambda_1, \lambda_2, \ldots, \lambda_n$ of $det(zI - A_d) = 0$ are called the poles of the digital system. The absolute values of stable poles are less than 1. If all the poles belong to stable region, the digital system is stable.

In the matrix $P(z) = \begin{bmatrix} A_d - zI & B_d \\ C_d & 0 \end{bmatrix}$, called the system matrix A, point $z = z_0$ satisfying $rank[P(z_0)] < min(n+m, n+l)$ is called an (invariant) zero. Note that n is the order, m is the number of inputs, and l is the number of outputs.

A.5.3 Reachability and Observability of Digital Systems

Definition A.5.2   Suppose we arbitrarily determine the initial value $x(0)$ and the final value x_e of the state variable $x(k)$. If, with some control input $u(k)$ and finite time $t_e(> 0)$, we can set $x(t_e) = x_e$, then we say the system (A_d, B_d) is **reachable**. In particular, when we can bring any $x(0) \neq 0$ to $x(t_e) = 0$, then we say that it is **controllable**. If $|A_d| \neq 0$, then they coincide.

---

12 $y(t) = Cx(t)$ to $y(k) = Cx(k)$ holds, but we have replaced it with $C_d = C$ in conjunction with A_d, B_d. In MATLAB, this can be calculated as $[Ad, Bd, Cd, Dd] = c2dm(A, B, C, D, T)$.

> **Theorem A.5.1** *It is equivalent for a system to be reachable and for the following to hold:*
> *(1) The rank of the reachability matrix* $U_c = [B_d, A_d B_d, A_d^2 B_d, \ldots, A_d^{n-1} B_d]$ *is n.*
> *(2) There exists a gain matrix* F_d *with arbitrary eigenvalues of* $A_d - B_d F_d$.

If the system is reachable, the state feedback $u(k) = -F_d x(k)$ allows all poles to be set arbitrarily, thus stabilizing the system.

> **Definition A.5.3** If the initial value $x(0)$ of the state variable is uniquely determined from the results of the control input $u(k)$ and the observation output $y(k)$ from $t = 0$ to time $t_e (> 0)$, then the system (A_d, C_d) is **observable**.

> **Theorem A.5.2** *It is equivalent for a system to be observable and for the following to hold:*
>
> *(1) The rank of the observability matrix* $U_o = \begin{bmatrix} C_d \\ C_d A_d \\ C_d A_d^2 \\ \vdots \\ C_d A_d^{n-1} \end{bmatrix}$ *is n.*
>
> *(2) There exists a matrix* H_d *with arbitrary eigenvalues of* $A_d - H_d C_d$.

A digital observer can be designed to estimate the state variable $x(k)$ if observable.

A.5.4 Digital State Feedback Control System Design

If (A_d, B_d) is controllable, then stabilization is possible with state feedback control. The control input is denoted as $u(k) = -F_d x(k)$. The system can be stabilized by $x(k+1) = A_d x(k) + B_d u(k) = A_d x(k) + B_d(-F_d x(k)) = (A_d - B_d F_d)x(k)$. The eigenvalues of $A_d - B_d F_d$ are called the poles of the digital control system, and the necessary and sufficient condition for a stable digital control system is that the absolute values of all the poles are less than 1, that is they are within a circle of radius 1 in the complex plane.

A.5.5 Digital Servo System Design

To design a digital servo system that follows a step reference value $r(k) = r$ under a step disturbance $d(k) = d$, an integrator must be positively inside the feedback system as the internal model. The integral $w(k + 1) = w(k) + Te(k)$ of the deviation $e(k) = r - y(k) = r - C_d x(k)$ between the reference value $r(k)$ and the output $y(k)$. However, some examples define the integral operation as $w(k + 1) = w(k) + e(k)$. The equation of state $x(k + 1) = A_d x(k) + B_d u(k) - d$ together with the integral $w(k)$ gives the expanded system $\bar{x}(k) = [x(k)^T, w(k)^T]^T$ and the following expanded system equation of state:

$$\begin{bmatrix} x(k + 1) \\ w(k + 1) \end{bmatrix} = \begin{bmatrix} A_d & 0 \\ -C_d T & I \end{bmatrix} \begin{bmatrix} x(k) \\ w(k) \end{bmatrix} + \begin{bmatrix} B_d \\ 0 \end{bmatrix} u(k) + \begin{bmatrix} -d \\ rT \end{bmatrix} \tag{A.55}$$

In the case of $t \to \infty$,

$$\begin{bmatrix} x(\infty) \\ w(\infty) \end{bmatrix} = \begin{bmatrix} A_d & 0 \\ -C_d T & I \end{bmatrix} \begin{bmatrix} x(\infty) \\ w(\infty) \end{bmatrix} + \begin{bmatrix} B_d \\ 0 \end{bmatrix} u(\infty) + \begin{bmatrix} -d \\ rT \end{bmatrix} \tag{A.56}$$

If the system can be stabilized, in $k \to \infty$, $[x(k + 1)^T, w(k)^T]^T = [x(k + 1)^T, w(k)^T]^T$, and the difference between Equations (A.55) and (A.56) is $e_{\bar{x}}(k) = [x(k)^T, w(k)^T]^T - [x(\infty)^T, w(\infty)^T]^T$. As $e_{\bar{u}}(k) = u(t) - u(\infty)$, subtracting the sides of Equations (A.55) and (A.56), we obtain Equation (A.57):

$$e_{\bar{x}}(k + 1) = \begin{bmatrix} A_d & 0 \\ -C_d T & I \end{bmatrix} e_{\bar{x}}(k) + \begin{bmatrix} B_d \\ 0 \end{bmatrix} e_{\bar{u}}(k) \tag{A.57}$$

Here assume we can design $e_{\bar{u}}(k) = -\bar{F}_d e_{\bar{x}}(k)$ which stabilizes Equation (A.57). Then we can get the following equation.[13]

$u(k) = u(\infty) - F_d e_{\bar{x}}(k) = u(\infty) - [F_d, -K_i]\{[x(k)^T, w(k)^T]^T - [x(\infty)^T, w(\infty)^T]^T\} = -F_d x(k) + K_i w(k) + \{u(\infty) + F_d x(\infty) - K_i w(\infty)\}$.

If $u(\infty) = -F_d x(\infty) + K_i w(\infty)$, then the control input $u(k) = -F_d x(k) + K_i w(k)$. Represented in the block diagram, this is Figure A.10.

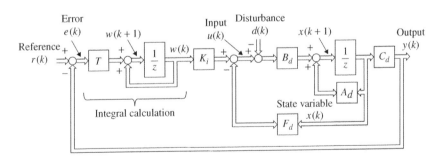

Figure A.10 Digital servo system block diagram.

---

13 We can find the adequate servo gains using the general pole assignment method (e.g. place.m) or discrete-time LQ control method (e.g. dlqr.m). The digital observer design is the same as the design method for continuous system observers, except that the poles are specified within a circle of radius 1.

A.6 Representation and Meaning of Optimal Programming

A.6.1 What Is Optimal Programming?

When using **optimal programming**, there are three cases such as expressed with no **constraints**, with the linear functions, and with nonlinear functions. Furthermore, their constraints are classified into those expressed in terms of equality (equality constraints) and the case of constraints expressed in terms of inequality (inequality constraints).

An optimal plan with inequality and equality constraints is written as follows: "Find x that minimizes $f(x)$ subject to $g_i(x) \leq 0, h_j(x) = 0$," etc. Its abbreviated form is expressed as "min.$f(x)$ s.t.$g_i(x) \leq 0, h_j(x) = 0$."

Find x that minimizes the objective function $f(x)$ under the constraints $g_i(x) \leq 0$ ($i = 1 \sim m$) and $h_j(x) = 0$ ($j = 1 \sim l$). However, $x = [x_1, x_2, \ldots, x_n]^T \in R^n$.

In the absence of constraints, this would be simply written as "Find x that minimizes the objective function $f(x)$."

A.6.2 fmincon Function

As preliminary knowledge, here is how to use the fmincon function.

---

fmincon Function Usage

One typical format should be shown as
$x = fmincon(fun, x0, A, B, A_{eq}, B_{eq}, LB, UB, nonlcon, options)$.
The constraints and maximum and minimum values are expressed as follows:

- Linear inequality constraint: $Ax \leq B$
- Linear equality constraints: $A_{eq}x = B_{eq}, A_{eq}$
- Nonlinear inequality constraint: $C(x) \leq 0$
- Nonlinear equality constraint, $C_{eq}(x) = 0$
- Minimum and maximum values: $LB \leq x \leq UB$

If no condition is specified, [] is used. The *fun* is the objective function and should be created as a separate file. In addition, the corresponding nonlinear constraints are described in the *nonlcon* file.

Next, an example of an optimization calculation using the fmincon function in MATLAB is given as follows.[14]

Example A.6.1 (Simple optimization example.) Under the conditions of $x_1^2/9 + x_2^2/4 - 1 \leq 0, x_1^2 - x_2 - 1 \leq 0, -10 \leq x_1 \leq 100$, and $0 \leq x_2 \leq 100$.

Solve the problem of minimizing the function $f(x) = 2x_1x_2$. The initial value of the optimization calculation is $x_1 = x_2 = 10$.

The constraints represent the shaded area in Figure A.11. They indicate finding the value among the x_1 and x_2 values in the area that minimizes the objective function. An example of solving using the fmincon function is shown in List A.1.

List A.1: Example of fmincon usage.

```
1   %------------------------------------------------
2   % Example: Optimization of 2 inequality constraint
3   % Objective: f(x)=2*x(1)+x(2)
4   % Inequality constraint: x1^2/9+x2^2/4-1<=0, and x1^2-x2-1<=0
5   %------------------------------------------------
6   %% Initial value and, for example, minimum value and maximum value
7   x0=[10;10];x_min=[-10;0];x_max=[100;100];
8
9   %% fmincon related number of experiments
10  [x,fval]=fmincon(@myfunc_opt,x0,[],[],[],[], x_min,x_max,@nonlcon)
11  % myfunc_opt is the destination number. nonlcon is a nonlinear constraint.
12  % x is the optimal value and fval is the minimum value of the destination number.
```

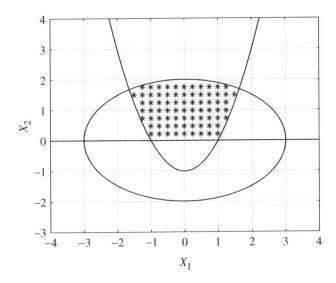

Figure A.11 Example of a range of constraints.

14 The name "fmincon" would mean minimizing function f under constraints.

```
x =

 −1. 0008e + 00
  1. 5094e + 03


fval =

 −2. 0000e + 00
```

Figure A.12 Example of fmincon function output.

Examples of the corresponding objective functions are also given in List A.2.

List A.2: Example of objective function in fmincon.

```
1  %−−−−−−−−−−−−−−−−−−−−−−−−−−−−−−
2  % The objective f(x)=2*x(1)+x(2)
3  %−−−−−−−−−−−−−−−−−−−−−−−−−−−−−−
4  function f=myfunc_opt(x)
5  f=2*x(1)+x(2);
```

An example of the description of nonlinear constraints is shown in List A.3.

List A.3: Example of nonlinear constraints in mincon.

```
1  %−−−−−−−−−−−−−−−−−−−−−−−−−−−−−−−−−−−−−−−−−−
2  % Inequality constraint conditions are described
3  %−−−−−−−−−−−−−−−−−−−−−−−−−−−−−−−−−−−−−−−−−−
4  function [c,ceq] = nonlcon(x)
5  c=[(x(1)^2)/9+(x(2)^2)/4−1;
6  x(1)^2 − x(2)−1];
7  ceq=[]; % this example is a nonlinear equation constraint []
8  end
```

The calculation results in optimal values of $x_1 = -1.0008$ and $x_2 = 0.0015$, and the minimum value of the function is -2.0. The execution result is the output as shown in Figure A.12.

A.6.3 Example of a Drawing Program

Using the Simulink model in Figure A.13 as an example, we show how to create a drawing waveform. Open the scope of the Simulink model and the configuration properties of Figure A.14a,b. Check the "Log data to workspace" at the check box. Give an appropriate variable name. In this example, the save format is "structure with time."

After executing the sim function from the m-file in MATLAB, a program such as List A.4 can be created and executed to obtain graphs such as Figure A.15a,b.

Figure A.13 An example of Simulink model.

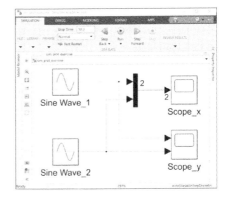

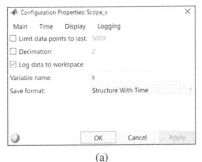

(a)

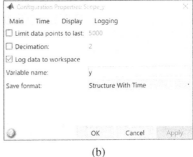

(b)

Figure A.14 Example of Scope settings for drawing Simulink data. (a) Scope x settings and (b) Scope y settings.

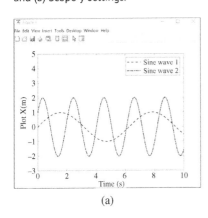

(a)

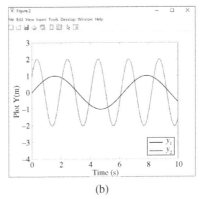

(b)

Figure A.15 Drawing example. (a) Scope x waveform and (b) Scope y waveform.

List A.4:   Example of a drawing program using the plot function.

```
1   %%%%%%%%%%%%%%%%%%%%%%
2   % Plot Example
3   %%%%%%%%%%%%%%%%%%%%%%
4   open('sim_plot_exercise');
5   set_param('sim_plot_exercise','WideLines','on'); % bold
6   set_param('sim_plot_exercise','ShowLineDimensions','on');% dimensional
7   z=sim('sim_plot_exercise');% simulink model run
8
9   %% Example of a drawing program using the Plot feature
10  figure(1);clf(1); tx=z.x.time; %Time component of log data
11  x1=z.x.signals.values(:,1);x2=z.x.signals.values(:,2);
12  plot(tx,x1,'--k',tx,x2,'-.k','LineWidth',1.2); grid on; % k black, -- is dashed line
13  xlabel('Time [s]','Fontsize',16);ylabel('Plot X[m]','Fontsize',16);
14  axis([0,10,-3,5]); % Set horizontal and vertical axis
15  legend('Sine wave 1','Sine wave 2','Fontsize',16);
16  set(gcf,'color','w'); % change window to white
17  set(gca,'Fontname','Times New Roman','FontSize',14); % Font setting
18
19  % Display of y
20  figure(2);clf(2);ty=z.y.time;y1=z.y.signals(1).values;y2=z.y.signals(2).values;
21  plot(ty,y1,'k',ty,y2,'k:','LineWidth',1.4);
22  xlabel('Time [s]','Fontsize',16);ylabel('Plot Y[m]','Fontsize',16); %Set label
23  axis([0,10,-4,3]);legend('y1','y2','Location','Southeast'); % legend, location
24  (Set up in the same way below)
```

Scope x and Scope y differ in whether the two signal lines are combined into one or connected as two in the MUX. The stored variables are handled differently depending on them.[15]

References

1 William Brogan: Modern Control Theory, Prentice Hall, 1990.
2 B. Gopinath: On the control of linear multiple input–output systems, The Bell System Technical Journal, Vol. 50, No. 3, 1063–1081, 1971.
3 Bruce A. Francis: A Cource in H_∞ Control Theory, Lecture Notes in Control and Information Sciences, Springer-Verlag, 1987.
4 Gene F. Franklin, Journal David Powell: Digital Control of Dynamic Systems, Addison Wesley, 1980.

---

15 When you want to create some figures with a plot function after executing Simulink, you should define any variable such as "z=" on the left-hand side of "sim ('Simulink model name')" in your program. However, this is unnecessary in the case of Simulink models created in older versions.

Index

Disturbance Observer for Advanced Motion Control with MATLAB/Simulink, First Edition. Akira Shimada.
© 2023 The Institute of Electrical and Electronics Engineers, Inc. Published 2023 by John Wiley & Sons, Inc.
Companion website: www.wiley.com/go/disturbanceobserver

Printed and bound by CPI Group (UK) Ltd, Croydon, CR0 4YY

16/04/2025

14658596-0002